Springer Tracts in Modern Physics

Volume 277

Springer Tracts in Modern Physics provides comprehensive and critical reviews of topics of current interest in physics. The following fields are emphasized:

- Elementary Particle Physics
- Condensed Matter Physics
- Light Matter Interaction
- Atomic and Molecular Physics
- Complex Systems
- Fundamental Astrophysics

Suitable reviews of other fields can also be accepted. The Editors encourage prospective authors to correspond with them in advance of submitting a manuscript. For reviews of topics belonging to the above mentioned fields, they should address the responsible Editor as listed in "Contact the Editors".

More information about this series at http://www.springer.com/series/426

Pierre Schnizer

Advanced Multipoles for Accelerator Magnets

Theoretical Analysis and Their Measurement

 Springer

Pierre Schnizer
Helmholtz-Zentrum Berlin für Materialien
   und Energie
Berlin
Germany

ISSN 0081-3869   ISSN 1615-0430   (electronic)
Springer Tracts in Modern Physics
ISBN 978-3-319-88077-8   ISBN 978-3-319-65666-3   (eBook)
DOI 10.1007/978-3-319-65666-3

Printed on acid-free paper

This Springer imprint is published by Springer Nature
The registered company is Springer International Publishing AG
The registered company address is: Gewerbestrasse 11, 6330 Cham, Switzerland

Acknowledgements

This treatise and the work described within this document were developed over the last years. Here, I wish to thank those without whom this treatise would not have been possible.

I am indebted to my colleague and supervisor, Egbert Fischer, who from the start has encouraged me to look aside of the straight path, to take the time to see the interesting topics along the road, yet continue to stay sufficiently focused such that the principal goal could be attained.

The study of the theoretical basis of this work started when its usefulness was only visible to those directly involved. It was the knowledge, experience and dedication of my father, Bernhard Schnizer, which allowed the development of these new tools that are now available to the community.

My sincere thanks are also directed to my mentors, Prof. Achim Denig, Prof. Frank Mass and Prof. Kurt Aulenbacher at the Johannes-Gutenberg-Universität Mainz, for providing me the chance and support to conduct this habilitation at the Institut für Kernphysik.

Experiments require a facility, planning, preparation and the support of many. I am indebted to all my colleagues at GSI and partner institutes, whose work only made it possible to build the measurement mole, to measure the magnets and which freed me from the stress of having to worry about all the particular subsystems required to run a superconducting magnet measurement bench successfully.

Last but not least, I wish to thank my wife Elisabeth for her faith in me, her support and patience and for accepting my absences related to this work over the last years.

Contents

1	**Introduction**	1
	1.1 Motivation	1
	1.2 The Laboratory and Its New Project	3
	1.3 The Magnets	4
	1.4 Scope of this Treatise	7
	References	9
2	**Electromagnetic Fields and Particle Motion**	11
	2.1 Maxwell's Equations	11
	2.1.1 Magnetic Quasistatic Approximation	12
	2.1.2 Magnetic Field in Linear Material	13
	2.2 Particle Motion in Magnetic Fields	17
	2.2.1 Particle Motion in a Cyclotron	18
	2.2.2 Paraxial Approximations	18
	2.2.3 Summary	19
	References	19
3	**Coordinate Systems**	21
	3.1 Orthogonal Curvilinear Systems	21
	3.2 Cylindrical Coordinate Systems	23
	3.2.1 Cylindrical Circular Systems	23
	3.2.2 Cylindrical Elliptical Systems	24
	3.3 Toroidal Coordinate Systems	27
	3.3.1 Global Toroidal Coordinates	27
	3.3.2 Local Toroidal Coordinates	27
	3.3.3 Local Toroidal Elliptical Coordinates	29
	3.4 Frenet-Serret Coordinates	31
	3.5 Summary	32
	References	32

4   Field Descriptions   35
   4.1   Basis: Cylindrical Circular Multipoles   36
       4.1.1   Conventions   37
       4.1.2   Effect of Transformations   40
       4.1.3   Circular Multipoles Using Real Variables   40
   4.2   Cylindrical Elliptical Multipoles   41
       4.2.1   Complex Elliptical Multipoles   43
       4.2.2   Relations Between Circular and Elliptical
               Multipoles   47
       4.2.3   Elliptical Multipole Field Expansions for Elliptical
               Components   51
       4.2.4   Complex Potential for Normal and Skew Elliptical
               Multipoles   53
   4.3   Toroidal Circular Multipoles   55
       4.3.1   Approximate R-Separation   56
       4.3.2   The Potential   57
       4.3.3   Real Basis Vector Field in Local Toroidal
               Coordinates   58
       4.3.4   Approximation Error of the Differential Equation   69
   4.4   Toroidal Elliptical Multipoles   69
   4.5   Summary   71
   References   72

5   Rotating Coils   75
   5.1   Derivation of Coil Probe Geometry Factors   75
       5.1.1   Complex Potential   76
       5.1.2   Magnetic Flux Through a Surface   76
       5.1.3   Magnetic Flux Picked Up by a Rotating Coil   78
   5.2   Radial Rotating Coil Layout   79
   5.3   Voltage Induced in a Rotating Pick Up Coil   80
   5.4   Compensated Systems   81
   References   84

6   Experimental Setup   85
   6.1   Test Facility   85
   6.2   The Anticryostat   87
   6.3   Magnetic Measurement Equipment   89
       6.3.1   History: Choice of Method   89
       6.3.2   A Modular Mole   91
   6.4   Summary   98
   References   99

7   Applications.   101
    7.1   Appropriate Handling of Calculation Data   101
    7.2   Calculating Cylindrical Elliptical Multipoles.   102
    7.3   Summary   105
    References.   106

8   Measuring Advanced Multipoles   107
    8.1   Measuring Straight Elliptical Multipoles.   108
          8.1.1   Calculation Procedure   108
          8.1.2   Measurement Results.   116
    8.2   Measuring Toroidal Multipoles.   116
          8.2.1   The Magnetic Flux   120
          8.2.2   Conversion Matrices   122
          8.2.3   Choosing a Coil Probe Length   124
          8.2.4   Magnitude of the Terms   126
          8.2.5   Measurement Results on the SIS100 Curved
                  Dipole Magnet   129
    8.3   Summary   130
    References.   130

9   Error Propagation.   133
    9.1   Error Propagation of Elliptic Multipoles Measurement   133
          9.1.1   Description of Calculation Procedure   133
          9.1.2   Combining the Coefficients   135
          9.1.3   Error Propagation of the Measured Coefficients   138
          9.1.4   Error Propagation to Coefficients of the Circular
                  Multipoles.   141
          9.1.5   Influence of Coil Probe Displacement.   144
    9.2   Toroidal Multipole Measurement   144
    9.3   Measurement Accuracy Estimate for the CSLD   145
    9.4   Summary   146
    References.   146

10 Conclusions   149
    10.1 Outlook.   151

Appendix A: Changes to Previous Publications   153

Appendix B: Mathematica Scripts.   155

Appendix C: Approximate Inversion of a Perturbed Matrix.   165

Abstract

Studying the transversal beam dynamic of accelerators requires a sound description of the magnetic field homogeneity. Cylindrical circular multipoles, typically used, have shortcomings if the beam aperture is elliptical or the curvature of the particle path has to be taken into account.

Within this treatise, advanced multipoles are described besides their application on numerical data and measurements. These multipoles allow describing the field consistently for elliptical and toroidal reference volumes and estimating the artefacts of the field deterioration. The higher precision of the field description gives better estimates of the beam dynamics and permits one to better estimate the effects of field inhomogeneity than was available before. Thus, the presentation itself does not create spurious artefacts; the margin normally reserved for these artefacts can be reduced due to the better understanding. This research was first necessitated by the R&D for developing the SIS100 magnets for FAIR, then used for describing the field homogeneity of the magnets next to developing appropriate measurement methods.

These findings are then further elaborated demonstrating their usefulness for describing the field of real magnets and their advantages over the cylindrical circular ones. Furthermore, measurement methods for obtaining the coefficients of these advanced multipoles have been developed and are described within this treatise. All these developments provided a basis for the field measurements of the first SIS100 model and first of series dipole magnets. The research results are then complemented with data obtained on the first SIS100 model and first of series dipole magnets.

Chapter 1
Introduction

1.1 Motivation

Accelerators have been at the core of large scale scientific instruments for driving our understanding of physics; starting with early forms of the "cathode ray tube" which allowed producing X-Rays and supported the discovery of the electron and the proton; similarly the Kaufmann-Bucherer-Neumann experiment set a first test on the theories explaining the relativistic mass and then later on to the theory of special relativity.

Soon the quest for higher energies turned table top experiments to setups filling halls as simple static accelerators were turned into cyclotrons, linear accelerators and synchrotrons. These ever larger installations were demanding larger investments leading to international collaborations with the most prominent ones the European Center for Nuclear Research (CERN) in Geneva and the Joint Institute for Nuclear Research (JINR) in Dubna.

The discovery of strong focusing [1] and its rediscovery [2–4] was the first time that the continuous increase of the machine size was not required anymore, as the same beam performance could be realised with a smaller machine. Furthermore a whole subscience emerged: accelerator physics, which focuses on studying and understanding the properties of accelerated particle beams.

As automatic computing machinery became affordable and the ready available power steadily increased, the developed physics description allowed implementing numerical models of the beam physics, which were bench marked with machine experiments. These efforts are, to a significant fraction, devoted to the movement of individual particles and particle bunches in the plane transversal to the beam propagation direction. The artifacts of the transversal particle movement are mainly dominated by imperfections of the used magnetic fields. Thus beam physic models require a solid description of these fields.

The magnet size is mainly defined by the used technology, the field strength, the aperture to be provided for the beam and the field quality required within this aperture. Particle beam accelerators for scientific research are typically large machines and of

© Springer International Publishing AG 2017
P. Schnizer, *Advanced Multipoles for Accelerator Magnets*, Springer Tracts
in Modern Physics 277, DOI 10.1007/978-3-319-65666-3_1

considerable investment; for circular machines the magnets and their supply systems represent a large share of this investment. Therefore any effort will be undertaken to reduce their aperture and thus the overall system size and so the cost. This is only feasible if the magnets and the artifacts, which they create, are thoroughly studied and the effects, which the artifacts provoke, are understood and controllable. Thus a consistent field description is required. For SIS100 (see Sect. 1.2) the commonly requested field quality of 200 ppm (i.e. the maximum tolerable deviation within the beam aperture) was increased to 600 ppm, which permitted designing compact magnets, necessary for realising them as compact superferric ones with cold iron. The increased field homogeneity target was seconded by a sound numerical study of the beam properties (e.g. beam stability, particle loss), which required field descriptions beyond the standard tool i.e. circular 2D multipoles. The higher field deviation to be tolerated is only possible due to the development of the advanced multipoles and their application. This reduces the uncertainty of the description. The purity of the magnetic field is less of a problem since the perturbations can be understood and their actions can be mitigated. The higher precision of the field description gives better estimates of the beam dynamics and permits one to better estimate the effects of the field inhomogeneity than was available before. These multipoles are based on elliptic and toroidal multipoles, which reflect the curved particle trajectory and the elliptic vacuum chamber and thus the beam aperture. These multipoles proved to be a reliable description and thus removed uncertainties in the field representation. Thus the machine designer is relieved from demanding a little better field quality than absolutely necessary foreseeing some extra margin for artifacts which were covered up by a less accurate field representation.

The developments presented in the following chapters were targeted to provide reliable field descriptions for the SIS100; here a beam area of elliptical cross section and curved trajectory has to be covered by a consistent description. Furthermore measurement methods for obtaining the coefficients of these advanced multipoles have been developed and are described within this treatise. The research results are complemented with coefficients of these advanced multipoles obtained from FEM models of the magnetic field and from measurements on the first SIS100 model magnets. Similar magnets have been used and described in beam dynamic calculations; in this treatise it is the first time that these descriptions are exact or approximate solutions of the potential equation. Approximate solutions are used for the toroidal geometries, as these are precise enough for the envisaged applications and more straightforward to treat and interpret than the equivalent exact solutions. The developments and methods given within this treatise have already found their application on studying the SIS100 machine performance and measuring its magnets fields. The described mathematics and measurement methods are, however, generally applicable for describing fields within cylindrical elliptical, toroidal cylindrical or toroidal elliptical geometry, as long as these fields are described by the potential equation.

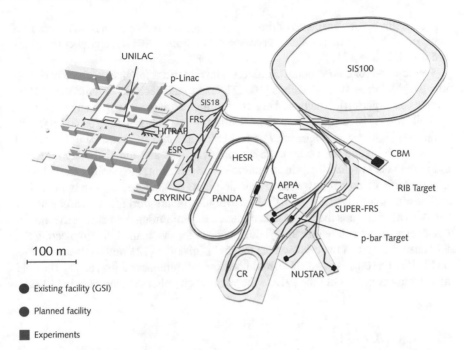

Fig. 1.1 The Facility for Antiproton and Ion Research. The accelerators already existing at GSI are indicated in *blue*. These are the universal linear accelerator *(UNILAC)* and the heavy ion synchrotron SIS18. All accelerators, which are part of the FAIR project, are depicted in *red*. Protons will be accelerated by a dedicated linear accelerator *(p-LINAC)*. The SIS100 synchrotron, the core component, will use fast ramped superconducting magnets. The super fragment separator *(SuperFRS)* is used to analyse rare isotopes. The collector ring (CR) will be used to cool and accumulate the antiprotons. The high energy storage ring *(HESR)* is foreseen for proton and antiproton experiments. Copyright: GSI/FAIR

1.2   The Laboratory and Its New Project

As indicated above, the multipole developments have their origin in the demands for the design and construction of the accelerator SIS100 as part of the FAIR project at GSI (see Fig. 1.1, [5]). This new **F**acility for **A**ntiproton and **I**on **R**esearch consists of several new accelerators, which use the existing universal linear accelerator (UNILAC) and the heavy ion synchrotron SIS18 as injectors. It will provide two synchrotrons dedicated for accelerating heavy ions, several fixed target experiments and various storage rings. An overview is given in [6].

The heavy ion synchrotron (German : "**S**chwer**I**onen**S**ynchrotron") with a beam rigidity $(B\rho)$ of 100 Tm is designed for high current high density beams of medium charge state ions. If these ions are scattered by residual gas atoms, potentially recombination effects can occur, which will result in beam loss. Therefore the beam has to be transported within vacuum of extreme quality (below 10^{-12} mbar) within the whole machine; for a machine of this size only considered achievable with vacuum chambers, which work as reliable cryo-vacuum pumps when the chamber temperature is well below 15 K [7–9].

The demand for cold vacuum chambers, and due to a reference machine in operation, the Nuclotron at JINR/Dubna [10, 11], lead to the decision to build the SIS100 magnets as superferric magnets, i.e. iron dominated magnets with a superconducting coil, cooled by a forced two phase Helium flow. The synchrotron SIS100 will be operated with cycle frequencies up to 1 Hz; thus large AC losses will be created in the iron yoke and the coil, due to hysteresis effects and induced eddy currents, and in the vacuum chamber, due to induced eddy currents. Recooling these losses requires large electrical power as cryogenic plants can theoretically only provide a cooling efficiency of $\approx 1.5\%$ with the best available plants working at an efficiency of $\approx 0.4\%$. This demands the losses to be reduced to a minimum, which in turn requires a magnet aperture solely providing enough space for the beam, but still meeting the field quality target. A overview on the magnet is given in [12], and its cooling limits in [13–15]. The efforts of AC loss minimisation are summarised in [16, 17], the field quality improvements in [18, 19] and the vacuum chamber development in [20–23].

1.3 The Magnets

A considerable part of this treatise is devoted to the theory required for deriving the advanced multipole descriptions. But, as it was motivated by the necessities of the FAIR project, the theoretical results had to be applied to the magnets, which were designed, built and tested for the SIS100 machine, the core component of the FAIR project.

The SIS100 main magnets are of the superferric type[1]: the magnets are iron dominated magnets, i.e. the field is formed by an iron yoke (see Fig. 1.2, a detail description is given in [12]). The magnet coil, however, is made of superconductors wound to a Nuclotron type cable: the superconductor strands are wrapped around a tube, which is cooled by a forced two phase Helium flow [11, 27]. The Lorentz force acts on the superconducting strands; a NiCr wire fixes them to the tube. Kapton is wrapped around them for electric insulation. The cable is wound to a coil, which is stiffened by a glass fibre reinforced epoxy. The coil support structure is precisely fabricated, to control the cable's position within the magnet yoke. The yoke is cooled to limit its heat leakage to the coil, so that the superconducting wires are below the critical temperature for the given operating conditions. The accelerator magnets of the same type and, in particular, the main dipoles and quadrupoles are typically connected electrically in series using bus bars. The bus bars for the dipole circuit and the 3 different quadrupole circuits are made of superconducting cable and are, thus, mounted on the magnet within its cryostat. The helium supply and return headers are mounted underneath the magnet yoke. The magnet is connected to the cryostat

---

[1] The SIS100 corrector magnets are, except for the chromaticity sextupole, air coil magnets [24–26].

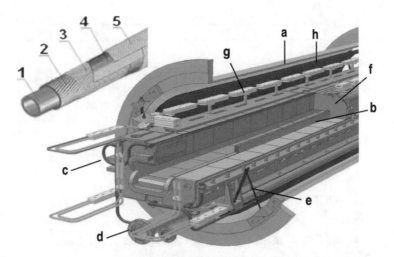

Fig. 1.2 The structure of the Nuclotron cable (inset *top left*) and the main features of the magnet design of the first full size dipole model. *1*—cooling tube, *2*—superconducting wire (multifilament NbTi/Cu), *3*—Nichrome wire, *4*—Kapton tape, *5*—adhesive Kapton tape, *a*—cryostat vessel, *b*—cable and half coil (2 · 4 windings), *c*—yoke cooling pipes, *d*—LHe lines, *e*—suspension rods, *f*—soft iron yoke, *g*—bus bars, *h*—thermal shield. Figure courtesy of Babcock Noell GmbH

with rods. The thermal shield of the cryostat allows reducing the heat leakage from the outside to the cold mass of the magnet.

Model dipole magnets were fabricated and two of them were tested and their fields were measured. Furthermore the series of SIS100 dipole magnets, which are curved and utilise a single layer coil, has been launched with the first of series magnet already delivered and currently being tested. These magnets are named as:

S2LD a straight dipole of 2.6 m length with a double layer coil and a vertical aperture of 68 mm [17, 28–33];

C2LD a curved dipole with a double layer coil, 3.1 m long with a vertical aperture of 66 mm, and a bending radius of 52.625 m [34, 35].

CSLD a curved dipole with a single layer coil, 3.1 m long with a vertical aperture of 68 mm and a bending radius of 52.625 m. Its development and its test results are summarised in [12, 15, 36].

The 2D cross section of the straight dipole is presented in Fig. 1.3 [37, 38] and a photo of the magnet in Fig. 1.4. A picture of the curved magnet is given in Fig. 1.5. The 2D cross section of the curved single layer dipole is given in Fig. 1.6 and a photo of the magnet is given in Fig. 1.7.

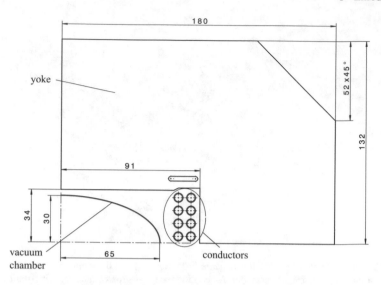

Fig. 1.3 Cross section of the SIS100 first full size dipole S2LD. All dimensions in millimetres

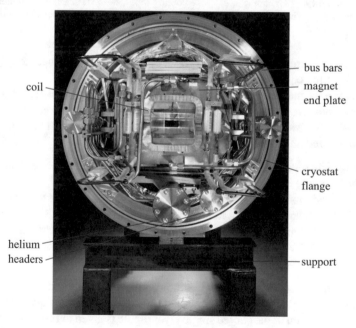

Fig. 1.4 Photo of the first full size straight dipole magnet (S2LD). Courtesy of Babcock Noell GmbH

Fig. 1.5 Photo of the first curved dipole magnet with a two layer coil (C2LD)

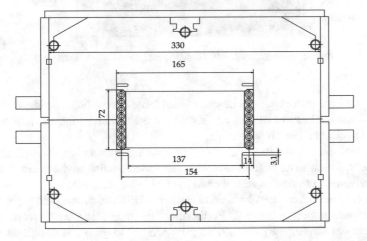

Fig. 1.6 Cross section of the SIS100 first of series curved single layer dipole (CSLD). All dimensions are rounded to millimetre

1.4   Scope of this Treatise

This document focuses on advanced multipole descriptions for magnetic fields, which were developed and were used to describe the design and measured fields of SIS100 magnets.

1. The treatise gives first Maxwell equations (Chap. 2) and shows that the potential equation can be used to describe the field in the aperture. Furthermore the theory of motion of a charged particle in magnetic fields is presented in a concise form.
2. Chapter 3 describes the basis of curvilinear coordinate systems together with various coordinate systems used within this treatise.

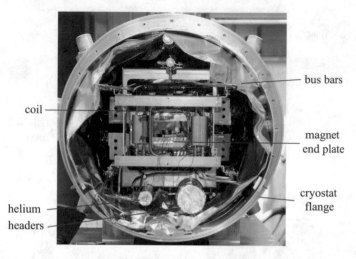

Fig. 1.7 Photo of first of series curved SIS100 dipole magnet (CSLD). Front view

3. Chapter 4 introduces various advanced multipole descriptions, details their properties and the basis functions. Furthermore it is shown how these can be recalculated to circular ones if applicable.
4. Chapter 5 introduces into the theory of rotating coil probes. The information
   presented in this chapter aids then later the understanding of the measurement of
   the advanced multipoles described in the following chapters.
5. Chapter 6 describes the experimental facility at GSI, which was used for operating
   the superconducting magnets. Furthermore the magnetic measurement system,
   which was developed for measuring the SIS100 magnets, is described and its
   peculiarities and challenges outlined.
6. The developments presented in Chap. 4 are now applied on calculated field data
   (see Chap. 7). The validity of this approach is demonstrated together with the
   usefulness of the tool.
7. The developed magnetic measurement system was used to measure the SIS100
   dipole magnets. A method was developed to construct elliptical cylindrical multipoles based on a set of rotating coil probe measurements. The results are presented
   together with the calculation procedure (see Chap. 8).
8. The descriptions of the magnetic field measurement are followed by an analysis
   of the error propagation, which allows deducing the measurement accuracy (see
   Chap. 9).
9. The results reflected in this treatise are summarised in Chap. 10.

References

1. N. Christofilos, *Focussing System for Ions and Electrons*, US Patent, Feb 1956
2. E.D. Courant, M.S. Livingston, H.S. Snyder, The strong-focusing synchroton—a new high energy accelerator. Phys. Rev. **88**, 1190–1196 (1952)
3. E.D. Courant, M.S. Livingston, H.S. Snyder, J.P. Blewett, Origin of the "strong-focusing" principle. Phys. Rev. **91**, 202–203 (1953)
4. E.D. Courant, H.S. Snyder, Theory of the alternating-gradient synchrotron. Ann. Phys. **3**, 1–48 (1958)
5. FAIR - Facility for Antiprotons and Ion Research, *Technical design report, synchrotron SIS100*, Dec 2008
6. FAIR - Facility for Antiprotons and Ion Research, *Technical design report,* Dec 2008
7. R.A. Haefer, *Cryopumping: Theory and Practice* (Clarendon Press, Monographs on Cryogenics, 1989)
8. V. Baglin, Cold/sticky systems, in *CAS—CERN Accelerator School—Vacuum in Accelerators*, ed. by D. Brandt (CERN, Geneva, 2007), pp. 351–368
9. S. Wilfert. *Investigations of the Dynamic Vacuum Conditions in the BINP SIS100 Dipole Vacuum Chamber During Magnet Ramping,* GSI Internal Note, Nov. (2011)
10. A.A. Smirnov, A.M. Baldin, A.M. Donyagin, E.I. D'yachkov, I.A. Eliseeva, H.G. Khodzhibagiyan, I.S. Khukhareva, A.D. Kovalenko, YuV Kulikov, B.K. Kuryatnikov, E.K. Kuryatnikov, V.N. Kuzichev, L.G. Makarov, P.I. Nikitaev, M.A. Voevodin, A.G. Zel'dovich, A.A. Vasiliev, A pulsed superconducting dipole magnet for the Nuclotron. Le J. de Phys. Colloq. **45**, 279–282 Jan (1984)
11. H.G. Khodzhibagiyan, A. Smirnov, The concept of a superconducting magnet system for the Nuclotron, in *Proceedings of the 12th International Cryogenic Engineerimg Conference ICIC12* (1988), pp 841–844
12. E. Fischer, P. Schnizer, *Design and Test Status of the SIS100 Dipole Magnet* (Report to the machine advisory committee, GSI Helmholtzzentrum für Schwerionenforschung mbH, Feb., 2010)
13. H.G. Khodzhibagiyan, A. Kovalenko, E. Fischer, Some aspects of cable design for fast cycling superconducting synchrotron magnets. IEEE T. Appl. Supercon. **14**(2), 1031–1034 (2004)
14. E. Fischer, H. Khodzhibagiyan, P. Schnizer, A. Bleile, Status of the SC magnets for the SIS100 synchrotron and the NICA project. IEEE T. Appl. Supercon. **23**(3), 4100504–4100504 (2013)
15. A. Bleile et al., Thermodynamic properties of the superconducting dipole magnet of the SIS100 synchrotron. *Physics Periodica*, vol. 67, 2015, Pages 1098–1101
16. E. Fischer, R. Kurnishov, P. Shcherbakov, Finite element calculations on detailed 3D models for the superferric main magnets of the FAIR SIS100 synchrotron. Cryogen. **47**, 583–594 (2007)
17. E. Fischer, P. Schnizer, A. Akishin, H. Khodzhibagiyan, A. Kovalenko, R. Kurnyshov, P. Shcherbakov, G. Sikler, W. Walter, Manufacturing of the first full size model of a SIS100 dipole magnet, in *WAMSDO Workshop*, number ISBN 978-92-9083-325-3, Jan (2009), CERN, pp. 147–156
18. P. Schnizer, B. Schnizer, P. Akishin, E. Fischer, Magnetic field analysis for superferric accelerator magnets using elliptic multipoles and its advantages. IEEE T. Appl. Supercon. **18**(2), 1605–1608 (2008)
19. P. Schnizer, B. Schnizer, P. Akishin, A. Mierau, E. Fischer, SIS100 dipole magnet optimisation and local toroidal multipoles. IEEE T. Appl. Supercon. **22**(3), 4001505–4001505 (2012)
20. S. Wilfert, K. Keutel, *Die kryogenen Vakuumkammern der supraleitenden magnete der SchwerIonenSynchrotron-Ringe SIS 100/300. Technical report,* Otto-von-Guericke-Universität (2004)
21. E. Fischer, P. Schnizer, C. Heil, A. Mierau, B. Schnizer, S. Shim. Impact of the beam pipe design on the operation parameters of the superconducting magnets for the SIS100 synchrotron of the FAIR project, in *12 European Conference on Cryogenics and Applied Superconductivity*. Journal of Physics: Conference Series, Sep (2009)

22. A. Mierau, P. Schnizer, E. Fischer, J. Macavei, S. Wilfert, S. Koch, T. Weiland, R. Kurnishov, P. Shcherbakov, Main design principles of the cold beam pipe in the fast ramped superconducting accelerator magnets for heavy ion synchrotron SIS100. Phys. Procedia **36**, 1354–1359 (2012)
23. A. Mierau, Numerische und experimentelle Untersuchungen gekoppelter elektromagnetischer und thermischer Felder in supraleitenden Beschleunigermagneten, PhD thesis, TU Darmstadt, Darmstadt, 2013
24. K. Sugita, E. Fischer, H. Khodzhibagiyan, H. Müller, J. Macavei, G. Moritz, Design of the multipole corrector magnets for SIS100. IEEE T. Appl. Supercon. **19**(3), 1154–1157 (2009)
25. K. Sugita, P. Akishin, E. Fischer, A. Mierau, P. Schnizer, 3D static and dynamic field quality calculations for superconducting SIS100 corrector magnets, in *Proceedings of IPAC'10*, May (2010), pp. 337–339
26. K. Sugita, E. Fischer, H. Khodzhibagiyan, J. Macavei, Design study of superconducting corrector magnets for SIS100. IEEE T. Appl. Supercon. **20**(3), 164–167 (2010)
27. N.N. Agapov, E.I. D'yachkov, H.G. Khodzhibagiyan, V.V. Krylov, YuV Kulikov, E.K. Kury-atnikov, V.N. Kuzichev, L.G. Makarov, N.M. Sazonov, A.A. Smirnov, V.V. Stekol'shchikov, A.G. Zel'dovich, A pulsed dipole magnet made from a hollow composite superconductor with a circulatory refrigeration system. Cryogen. **20**(6), 345–348 (1980). June 1980
28. G. Sikler et al., Full size model manufacturing and advanced design status of the SIS100 main magnets. WAMSDO at CERN, June (2008)
29. E. Fischer, H. Khodzhibagiyan, A. Kovalenko, P. Schnizer, Fast ramped superferric prototypes and conclusions for the final design of the SIS100 main magnets. IEEE T. Appl. Supercon. **19**(3), 1087–1091 (2009)
30. E. Fischer, P. Schnizer, R. Kurnyshov, B. Schnizer, P. Shcherbakov, Numerical analysis of the operation parameters of fast cycling superconducing magnets. IEEE T. Appl. Supercon. **19**(3), 1266–1267 (2009)
31. E. Fischer, P. Schnizer, P. Akishin, R. Kurnyshov, A. Mierau, B. Schnizer, P. Shcherbakov. Measured and calculated field properties of the SIS100 magnets described using elliptic and toroidal multipoles, in *PAC 09, Vancouver 2009,* May (2009)
32. E. Fischer, A. Mierau, P. Schnizer, C. Schroeder, A. Bleile, E. Floch, J. Macavei, A. Stafiniak, F. Walter, G. Sikler, W. Gärtner. Fast ramped superferric prototype magnets of the FAIR project, first test results and design update, in *PAC 09, Vancouver 2009,* May (2009)
33. G. Sikler, W. Gärtner, A. Wessner, E. Fischer, E. Floch, J. Macavei, P. Schnizer, C. Schroeder, F. Walter, D. Krämer, Fabrication of a prototype of a fast cycling superferric dipole magnet. In *PAC 09, Vancouver 2009,* May (2009)
34. E. Fischer, P. Schnizer, P. Akishin, R. Kurnyshov, A. Mierau, B. Schnizer, S.Y. Shim, P. Sherbakov, Superconducting SIS100 prototype magnets design, test results and final design issues. IEEE T. Appl. Supercon. **20**(3), 218–221 (2010)
35. P. Schnizer, E. Fischer, H. Kiesewetter, F. Klos, T. Knapp, T. Mack, A. Mierau, B. Schnizer, Commissioning of the mole for measuring SIS100 magnets and first test results. IEEE T. Appl. Supercon. **20**(3), 1977–1980 (2010)
36. E. Fischer, P. Schnizer, K. Sugita, J.P. Meier, A. Mierau, A. Bleile, P. Szwangruber, H. Müller, C. Roux, Fast ramped superconducting magnets for FAIR—production status and first test results. IEEE T. Appl. Supercon **25**(3), 1–5 (2015). Art.Nr:4003805
37. A. Kalimov, *SIS-100 dipole magnet. Technical report,* Gesellschaft für Schwerionenforschung, Planckstrasse 1, 64291 (Germany, Feb 2007)
38. E. Fischer, J. Macavei, A. Mierau, P. Schnizer, *The straight SIS100 dipole S2LD (Parameters and calculations. Technical report, GSI Helmholtzzentrum für Schwerionenforschung mbH),* (2008)

Chapter 2
Electromagnetic Fields and Particle Motion

The design of accelerator magnets requires an understanding of the generation of magnetic fields together with the motion of particles within these fields. The basic equations are presented here to clarify the utilised symbols. Maxwell's equations and the magneto quasistatic approximation (MQS) is presented in [1, 2]. The equations of particle motion is based on analytic mechanics.

2.1 Maxwell's Equations

The field descriptions presented in this treatise are based on Maxwell's equations. These are given in the differential form by

$$\nabla \cdot \mathcal{B} = 0 \tag{2.1}$$

$$\nabla \times \mathcal{H} = \mathcal{J} + \frac{\partial \mathcal{D}}{\partial t} \tag{2.2}$$

$$\nabla \times \mathcal{E} = -\frac{\partial \mathcal{B}}{\partial t} \tag{2.3}$$

$$\nabla \cdot \mathcal{D} = \rho_e, \tag{2.4}$$

with $\mathcal{B}$ the magnetic induction, $\mathcal{H}$ the magnetising field, t the time, $\mathcal{J}$ the electric current density, $\mathcal{D}$ the electric displacement field, $\mathcal{E}$ the electric field and ρ_e the electric charge density. Calligraphic letters denote vector fields. Any stationary material is modelled by

$$\mathcal{B} = \mu_0 \left(\mathcal{H} + \mathcal{M}(\mathcal{H}) \right) \tag{2.5}$$

$$\mathcal{D} = \epsilon_0 \mathcal{E} + \mathcal{P}_e(\mathcal{E}), \tag{2.6}$$

with $\mathcal{M}$ the magnetic polarisation and $\mathcal{P}_e$ the electric polarisation.

© Springer International Publishing AG 2017
P. Schnizer, *Advanced Multipoles for Accelerator Magnets*, Springer Tracts in Modern Physics 277, DOI 10.1007/978-3-319-65666-3_2

2.1.1   *Magnetic Quasistatic Approximation*

Here only magnetic fields are treated; electric fields only need to be considered as so far as they are having an impact on the magnetic fields. Given that field changes are slow the time dependence of the term $\frac{\partial \mathcal{D}}{\partial t}$ can be considered not significant within the domain of interest, so the above equations reduce to

$$\nabla \cdot \mathcal{B} = 0 \tag{2.7}$$

and the magnetising $\mathcal{H}$-field is only induced by the current

$$\nabla \times \mathcal{H} = \mathcal{J} + \underbrace{\frac{\partial \mathcal{D}}{\partial t}}_{\approx 0} \approx \mathcal{J}. \tag{2.8}$$

This equation implies

$$\nabla \cdot \mathcal{J} = 0, \tag{2.9}$$

as no electric currents nor electric displacement currents are considered within the magneto quasi-static approximation (MQS).

   If the current density vanishes, $\mathcal{J} = 0$, then from (2.8) follows (e.g. [1])

$$\nabla \times \mathcal{H} = 0, \tag{2.10}$$

which implies that a potential can be defined by

$$\mathcal{H} = -\nabla \Phi'. \tag{2.11}$$

In the absence of material $\mathcal{B} = \mu_0 \mathcal{H}$. Inserting this equation in (2.1) leads to

$$\nabla \cdot \mu_0 \mathcal{H} = 0, \tag{2.12}$$

and the Laplace equation for the scalar magnetic potential Φ',

$$\Delta \Phi' = 0 \tag{2.13}$$

is obtained. As this treatise focuses on fields in air gaps, it will relate any further potential to the magnetic induction $\mathcal{B} = \mu_0 \mathcal{H} = -\nabla \Phi$.

2.1.2 Magnetic Field in Linear Material

The magnetic material Eq. (2.5) shows two components. The first one describes the relation of the magnetising field to the magnetic field in vacuum. Material, which is isotropic and magnetically linear, can be modelled by

$$\mathcal{M} = \mu_r \mathcal{H} \tag{2.14}$$

or more commonly by

$$\mathcal{B} = \mu_0 \mu_r \mathcal{H}, \tag{2.15}$$

with μ_r the scalar magnetisation. For theoretical considerations one writes

$$\mathcal{B} = \mu \mathcal{H}. \tag{2.16}$$

This μ is used to describe "linear material". Any material reacting to the magnetising field in an isotropic manner and independent of the strength of the magnetising field fulfils this property. Vacuum and air are two common practical materials of this category. For vacuum $\mu_r = 1$ while for air $\mu_r \approx 1 + 0.4 \times 10^{-6}$. The fields are treated here with an accuracy of not more than 1 ppm, thus the difference between air and vacuum can be neglected. The characteristics of linear material allow a further simplification of Maxwell's equations.

2.1.2.1 Poisson and Laplace Equation for the Vector Potential

Given that magnetic fields are solenoidal (see (2.1)) and using the assumption that only fields in linear material need to be considered here, one can rewrite (2.1) by

$$\nabla \cdot \mathcal{B} = \nabla \cdot \mu_0 \mathcal{H}. \tag{2.17}$$

A vector potential $\mathcal{A}$ can now be introduced by

$$\nabla \times \mathcal{A} = \mu_0 \mathcal{H}. \tag{2.18}$$

From (2.8), one can deduce that

$$\nabla \times (\nabla \times \mathcal{A}) = \mu_0 \mathcal{J}. \tag{2.19}$$

The above operator is recalculated to

$$\nabla \times (\nabla \times \mathcal{A}) = \nabla(\nabla \cdot \mathcal{A}) - \nabla^2 \mathcal{A} \tag{2.20}$$

following vector analysis rules. However, the vector potential is not unique. It may
be defined as purely solenoidal: $\nabla \cdot A = 0$ (Coulomb gauge) which in turn yields
$\nabla(\nabla{\cdot}A) = 0$. Thus one obtains an equation resembling the Poisson equation for the
vector potential

$$\Delta A = -\mu_0 \mathcal{J}. \tag{2.21}$$

Many problems are constant with respect to one of the coordinates; then only a
xy- plane needs to be considered with currents flowing only perpendicular to this
plane. If the third coordinate is a Cartesian one, say z, the above equation simplifies
to

$$\Delta A_z = -\mu_0 J_z . \tag{2.22}$$

In the free aperture of the magnet no currents or charges are found. The field is
described by the homogeneous equation corresponding to the above equations. But
in this case it is more advantageous to use a scalar magnetic potential Φ, which is a
solution of the potential equation

$$\Delta \Phi = 0 . \tag{2.23}$$

The derivation of this equation from Maxwell's equation was given in Sect. 2.1.1.

2.1.2.2 Solutions of the Laplace Equation

The solution of the Laplace Equation (2.13) depends, as for any partial differential
equation, on the boundary conditions next to the domain. Furthermore, the set of
functions to be used depends on the coordinate system (see e.g. [3]). Here they are
given for a circular cylinder. Solutions for other coordinate systems are given in
Chap. 4. The Laplace Equation in cylindrical circular coordinates is given by

$$\Delta \Phi = \frac{\partial^2 \Phi}{\partial r^2} + \frac{1}{r}\frac{\partial \Phi}{\partial r} + \frac{1}{r^2}\frac{\partial^2 \Phi}{\partial \vartheta^2} = 0, \tag{2.24}$$

assuming that the problem is independent of z, and thus, $\frac{\partial^2 \Phi}{\partial z^2} = 0$. This equation can
be solved by separation using $\Phi = R(r)\Theta(\vartheta)$ (see e.g. [3]), forming the differential
equations

$$\frac{d^2 R}{dr^2} + \frac{1}{r}\frac{dR}{dr} - \frac{a_2}{r^2} = 0, \quad \text{and} \tag{2.25}$$

$$\frac{d^2\Theta}{d\vartheta^2} + a_2\Theta = 0. \tag{2.26}$$

For $a_2 = 0$ the solutions are

$$R = A_0 + B_0 \ln(r) \tag{2.27}$$

$$\Theta = C_0 + D_0 \vartheta \,. \tag{2.28}$$

For $a_2 = n^2$ the general solution is given by

$$R = A'_n r^n + B'_n r^{-n} \tag{2.29}$$

$$\Theta = F'_n \sin(n\vartheta) + G'_n \cos(n\vartheta) \,. \tag{2.30}$$

Fields considered within this treatise are always within a reference volume; in this section it is a circular cylinder. A regular solution within this boundary is demanded, thus only the following functions are considered:

$$\sin(n\vartheta), \ \cos(n\vartheta), \ r^n \,. \tag{2.31}$$

Hence the potential Φ is given by

$$\Phi(r, \vartheta) = -\sum_{n=1}^{\infty} r^n \left(F_n \sin(n\vartheta) + G_n \cos(n\vartheta)\right) \,. \tag{2.32}$$

The associated field components are given by

$$B_r(r, \vartheta) = -\frac{\partial \Phi}{\partial r} = \sum_{n=1}^{\infty} n \, r^{n-1} \left(F_n \sin(n\vartheta) + G_n \cos(n\vartheta)\right) \quad \text{and} \quad \tag{2.33}$$

$$B_\vartheta(r, \vartheta) = -\frac{1}{r}\frac{\partial \Phi}{\partial \vartheta} = \sum_{n=1}^{\infty} n \, r^{n-1} \left(F_n \cos(n\vartheta) - G_n \sin(n\vartheta)\right) \,. \tag{2.34}$$

Now these field components, given in the cylindrical circular coordinate system shall be converted into a Cartesian coordinate system represented by complex coordinates. The Cartesian components are given by

$$B_y = B_r \sin\vartheta + B_\vartheta \cos\vartheta \quad \text{and} \quad B_x = B_r \cos\vartheta - B_\vartheta \sin\vartheta \,. \tag{2.35}$$

The geometric extent of magnetic fields, found in accelerators, is typically much longer in one coordinate than in the other two. The 3D field can be treated by beam optics as an infinite thin object (equivalent to a thin lens in geometric optics). Complex representation simplifies the calculation in 2D space. In 3D Cartesian coordinates Maxwell's equations of the magnetic field in linear material are described by

$$\nabla \cdot \mathcal{B} = \left(\frac{\partial B_x}{\partial x} + \frac{\partial B_y}{\partial y} + \frac{\partial B_z}{\partial z}\right) = 0 \tag{2.36}$$

and without currents within the reference volume by

$$\nabla \times \mathcal{B} = \begin{pmatrix} \frac{\partial B_z}{\partial y} - \frac{\partial B_y}{\partial z} \\ -\frac{\partial B_z}{\partial x} + \frac{\partial B_x}{\partial z} \\ \frac{\partial B_y}{\partial x} - \frac{\partial B_x}{\partial y} \end{pmatrix} = 0 . \tag{2.37}$$

As the problem is to be independent of the longitudinal coordinate z one can set $\frac{\partial B_y}{\partial z} = 0$, $\frac{\partial B_x}{\partial z} = 0$. Furthermore the field component B_z shall be invariant with respect to z, thus $\frac{\partial B_z}{\partial x} = 0$ and $\frac{\partial B_z}{\partial y} = 0$. This simplifies the above equations to

$$\frac{\partial B_x}{\partial x} = -\frac{\partial B_y}{\partial y} \quad \text{and} \quad \frac{\partial B_y}{\partial x} = \frac{\partial B_x}{\partial y} . \tag{2.38}$$

These are exactly the Cauchy-Riemann equations for $B_y + i B_x$ with $x + iy$. Using complex notation (2.35) can be expressed by [4]

$$B_y + i B_x = (B_\vartheta + i B_r) e^{-i\vartheta} . \tag{2.39}$$

$B_\vartheta + i B_r$ are thus (using (2.33) and (2.34))

$$B_\vartheta + i B_r = \sum_{n=1}^{\infty} n r^{n-1} (F_n + i G_n) e^{in\vartheta} . \tag{2.40}$$

Furthermore it is good practice to use dimensionless coefficients. Therefore the radius is scaled by

$$\rho = \frac{r}{R_{\text{Ref}}} , \tag{2.41}$$

with R_{Ref} the reference radius. Using the relation between the polar and Cartesian field (2.39) and the expansion (2.40) the complex field $\mathbf{B}(\mathbf{z})$ is given by

$$\mathbf{B}(\mathbf{z}) = B_y(x + iy) + i B_x(x + iy) = \sum_{n=1}^{\infty} \mathbf{C_n} \left(\frac{\mathbf{z}}{R_{\text{Ref}}} \right)^{n-1} , \tag{2.42}$$

with $\mathbf{C_n} = B_n + A_n$, $B_n = n F_n R_{\text{Ref}}^{n-1}$, and $A_n = n G_n R_{\text{Ref}}^{n-1}$. The B_n are called the normal components and the A_n the skew components. $\mathbf{C_n}$, B_n and A_n are called "absolute harmonics". Their dimension is T at R_{Ref} (Tesla at reference radius). For accelerators the so called normalised harmonics $\mathbf{c_n} = b_n + i a_n$ are introduced by

$$\mathbf{c_n} = \frac{\mathbf{C_n}}{\mathbf{C_m}} 10^4 , \tag{2.43}$$

with $\mathbf{C_m}$ the main harmonic of the magnet. The main harmonic is $\mathbf{C_1}$ for a dipole, $\mathbf{C_2}$ for a quadrupole, and so on. These coefficients are dimensionless. The scaling factor 10^4 is used so that the size of $\mathbf{c_n}$ is in the order of 1 for a typical accelerator magnet. In the accelerator community one then typically speaks of a "harmonic of the size of 1 unit".

2.1.2.3 Treating Fringe Fields as Averaged 2D Fields

Also fringe fields can be treated as 2D fields, if the 3D field is integrated over the longitudinal coordinate z [5]. An averaged potential $\overline{\Phi}$ is introduced, which is obtained integrating Laplace's equation in z over a distance z_L. Then for this case the equation can be reformulated to

$$\frac{\partial^2 \overline{\Phi}}{\partial x^2} + \frac{\partial^2 \overline{\Phi}}{\partial y^2} + \frac{1}{z_L} \frac{\partial \Phi}{\partial z}\bigg|_{z_0}^{z_0+z_L} = 0, \qquad (2.44)$$

with z_0 the integration limit. The field $\mathcal{B}$ is obtained by $-\nabla \Phi$. So $\frac{\partial \Phi}{\partial z}$ corresponds to B_z. Then the last term vanishes if B_z fulfils

$$B_z(z_0) = 0 \quad \text{and} \quad B_z(z_0 + L) = 0. \qquad (2.45)$$

So end fields can be treated as averaged 2D fields if these conditions are fulfilled.

2.2 Particle Motion in Magnetic Fields

Particles guided in accelerators, typically electrons, protons or ions, are charged. Due to their speed, $\vec{v}$, often between 10% to closely to 100% of the speed of light c, the Lorentz Force

$$\vec{F}_{\mathrm{L}} = q\,\vec{v} \times \mathcal{B} \qquad (2.46)$$

is much larger for technically achievable magnetic fields (typically between 0.3 to 2 T for conventional magnets) than the force on the charge q in a technical electric field

$$\vec{F} = q\,\mathcal{E}. \qquad (2.47)$$

Magnetic fields can not increase the particle energy but only deviate their paths due to the cross product in (2.46). Hence the particle motion considered here is limited only to deflections of the particle path. Such a deflection requires a centripetal force F_{zp}. Its strength is given by

$$\left| F_{\mathrm{zp}} \right| = \frac{m\,\left| v^2 \right|}{R} \qquad (2.48)$$

with m the mass of the particle, v its velocity and R the orbit's radius.

2.2.1 Particle Motion in a Cyclotron

Cyclotrons are accelerators and the particles gain energy within the gap of its electrodes. The treatise focuses on magnetic fields and thus only on particles with constant energy; therefore the energy gain of the particles is neglected. So here a cyclotron can be considered as a dipole magnet whose field area is so large that the whole particle path is covered. The equilibrium path is given if the forces are balanced by

$$\frac{m v_s^2}{r} = F_{zp} = F_L = q v_s B_y,$$ (2.49)

with v_s the particle velocity in direction of the particle path s. Using the relation $p = m v_s$ yields

$$\frac{1}{R} = \frac{q}{p} B_y.$$ (2.50)

If the field of the cyclotron is not totally uniform one obtains as local equilibrium

$$\frac{1}{R(x, y, s)} = \frac{q}{p} B_y(x, y, s).$$ (2.51)

2.2.2 Paraxial Approximations

Amongst other things the path of a particle in a cyclotron is defined by the particle velocity. Its local radius depends on the particular position (see (2.51)) [6, 7]. The relation (2.51) is generally valid. In synchrotrons or storage rings the magnetic field strength is adjusted to the particle's momentum, because otherwise the particles will hit the aperture and get lost.

The ideal orbit is the trajectory of the particle for which the accelerator is designed. Most of the particles in a bunch do not have the initial conditions for this orbit; but one assumes that their deviations from the ideal orbit are small. Different particles in an accelerator bunch are assumed to have only small deviations from the ideal orbit. Therefore one can study the field influence considering only local derivatives. Further the magnet fields are assumed to be independent of the longitudinal coordinate s.

Using a coordinate system, which follows the ideal particle path (i.e. Frenet-Serret coordinates, see Sect. 3.4) one can substitute $r = R + x$, with R the radius of the foreseen orbit of the ideal particle and x the horizontal offset from this ideal orbit. Then the force F_x is given by

$$F_x(r) = \frac{m v^2}{\underbrace{R + x}_{=r}} - e v B_y(R + x).$$ (2.52)

Given that the local field change is small, the field variation can be described by

$$B_y(r) = B_{y0} + \frac{\partial B_y}{\partial x} x + \frac{\partial^2 B_y}{\partial x^2} x^2 + \dots \qquad (2.53)$$

This allows describing the dependence of the field on the offset of the particle from the ideal orbit using

$$B_y(x) = B_{y0} \left[1 + \frac{R}{B_{y0}} \frac{\partial B_y}{\partial x} \frac{x}{R} \right]. \qquad (2.54)$$

In textbooks on beam dynamics (e.g. [7]) the derivatives in (2.53) are approximated using the coefficients of the power series (2.42) [4]:

$$\left. \frac{\partial^{n-1} B_y}{\partial x^{n-1}} \right|_{x=0,\, y=0} = \frac{(n-1)!}{R_{\text{Ref}}^{n-1}} \operatorname{Re}[\mathbf{C_n}] = \frac{(n-1)!}{R_{\text{Ref}}^{n-1}} B_n, \qquad (2.55)$$

$$\left. \frac{\partial^{n-1} B_x}{\partial x^{n-1}} \right|_{x=0,\, y=0} = \frac{(n-1)!}{R_{\text{Ref}}^{n-1}} \operatorname{Im}[\mathbf{C_n}] = \frac{(n-1)!}{R_{\text{Ref}}^{n-1}} A_n. \qquad (2.56)$$

This approach neglects the curvature of the particle trajectory within e.g. a long dipole magnet as the whole theory assumes that the "lenses" (i.e. the magnets) are short, similar to the "thin lens" approximation in geometric optics.

2.2.3   Summary

Particle accelerators use magnetic fields to guide the particles around the ideal path. The aperture of these magnets is evacuated, thus material free, and the ramp rates of the magnetic fields are small enough so that the magnetic quasistatic approximation can be used. As the aperture is material and current free, the potential of the magnetic field is described by Laplace equation (2.13).

The transverse offset of the particles from the ideal orbit is small, therefore beam dynamics of synchrotrons and colliders uses the paraxial approximation to describe the effect of the magnetic fields on the particle beam. Commonly approximate circular multipoles are used for that purpose, even if the particle path is significantly bent within the magnet (e.g. in a dipole in a small synchrotron).

References

1. H.A. Haus, J.R. Melcher, *Electromagnetic Fields and Energy* (Prentice-Hall Inc., New Jersey, 1989)
2. H. Hofmann, *Das elektromagnetische Feld. Theorie und grundlegende Anwendungen* (Springer, Wien, 1986)

3. P. Moon, *Field Theory Handbook: Including Coordinate Systems, Differential Equations and Their Solutions* (Springer, Berlin, 1988)
4. A. Wolski, Maxwell's equations for magnets, in *CERN Accelerator School: Specialised Course on Magnets, volume CERN-2010-004* ed. by D. Brandt, CERN, published as CERN Yellow Report, http://cdsweb.cern.ch/record/1158462 (2010), pp. 1–38
5. W.C. Elmore, M.W. Garrett, Measurement of two-dimensional fields. Part I: Theor. Rev. Sci. Instr. **25**(5), 480–485 (1954)
6. K. Wille, *Physik der Teilchenbeschleuniger und Synchrotronstrahlungsquellen* (Teubner, Wiesbaden, 1992)
7. H. Wiedemann, *Particle Accelerator Physics* (Springer, Berlin, 2007)

Chapter 3
Coordinate Systems

For many problems an adapted coordinate system allows reducing the number of dimensions if chosen appropriately. A familiar example is the use of the polar coordinate system, which simplifies any circular problem independent of the polar angle. Here coordinate systems are presented, which are commonly known but have not been frequently applied for describing the fields of accelerator magnets.

The major part of the developments presented below was made and is used for the design and numerical investigation of the SIS100 synchrotron. The vacuum chamber of these magnets are elliptical and the magnets are curved. These two features require

- a coordinate system which allows describing the field within the elliptical aperture and
- a coordinate system which follows the curvature of the magnet.

The field within the magnet can be described using Cartesian coordinates. Then, however, the field variation in the longitudinal direction can not be neglected, and thus the number of dimensions can not be reduced.

Many 3D problems can be reduced by one dimension as these problems are invariant versus one of the coordinates, if the appropriate coordinate system is chosen. A concise overview over these coordinate systems is given in [1]. Here two coordinate systems are given, which can support such an simplification: cylindrical systems and then toroidal systems.

3.1 Orthogonal Curvilinear Systems

A common curvilinear system u_i can be defined by (e.g. [2])

$$x_i = x_i(u_1, u_2, u_3) \quad \text{and} \quad \vec{x} = \vec{x}(u_1, u_2, u_3), \tag{3.1}$$

© Springer International Publishing AG 2017
P. Schnizer, *Advanced Multipoles for Accelerator Magnets*, Springer Tracts in Modern Physics 277, DOI 10.1007/978-3-319-65666-3_3

with x_i the coordinates of a Cartesian coordinate system: $x_1 = x$, $x_2 = y$, $x_3 = z$. The functions x_i have to be unique and continuously differentiable. Solving for u_i gives

$$u_i = u_i(x, y, z) \quad \text{and} \quad \vec{u} = \vec{u}(x, y, z). \tag{3.2}$$

Inserting (3.2) in (3.1) yields

$$x_i = x_i(u_1(x_1, x_2, x_3), \ u_2(x_1, x_2, x_3), \ u_3(x_1, x_2, x_3)). \tag{3.3}$$

Inserting (3.1) in (3.2) gives

$$u_i = u_i(x_1(u_1, u_2, u_3), \ x_2(u_1, u_2, u_3), \ x_3(u_1, u_2, u_3)). \tag{3.4}$$

Using $\partial x_i / \partial x_k = \delta_{ik}$ the differentiation of (3.3) has the properties

$$\frac{\partial x_i}{\partial x_k} = \frac{\partial x_i}{\partial u_l} \frac{\partial u_l}{\partial x_k} = \delta_{ik}. \tag{3.5}$$

A tangent $\vec{q}$ to the curve $\vec{x}(u)$ is given by

$$\vec{q} = \frac{\frac{\partial \vec{x}}{\partial u}}{\sqrt{\frac{\partial \vec{x}}{\partial u} \frac{\partial \vec{x}}{\partial u}}}. \tag{3.6}$$

The metric coefficients h_i of a orthogonal curvilinear system are given by

$$h_1^2 = \left(\frac{\partial x_1}{\partial u_1}\right)^2 + \left(\frac{\partial x_2}{\partial u_1}\right)^2 + \left(\frac{\partial x_3}{\partial u_1}\right)^2,$$

$$h_2^2 = \left(\frac{\partial x_1}{\partial u_2}\right)^2 + \left(\frac{\partial x_2}{\partial u_2}\right)^2 + \left(\frac{\partial x_3}{\partial u_2}\right)^2, \tag{3.7}$$

$$h_3^2 = \left(\frac{\partial x_1}{\partial u_3}\right)^2 + \left(\frac{\partial x_2}{\partial u_3}\right)^2 + \left(\frac{\partial x_3}{\partial u_3}\right)^2.$$

The gradient of a scalar field is given by $\nabla \mathcal{A}$

$$\{\nabla \mathcal{A}\}_{u_1} = \frac{1}{h_1} \frac{\partial \mathcal{A}}{\partial u_1}, \quad \{\nabla \mathcal{A}\}_{u_2} = \frac{1}{h_2} \frac{\partial \mathcal{A}}{\partial u_2}, \quad \{\nabla \mathcal{A}\}_{u_3} = \frac{1}{h_3} \frac{\partial \mathcal{A}}{\partial u_3}. \tag{3.8}$$

The scalar Laplace operator $\Delta = \nabla \cdot \nabla$ for a scalar field $\mathcal{A}$ is given by

$$\Delta \mathcal{A} = \frac{1}{h_1 h_2 h_3} \left[\frac{\partial}{\partial u_1} \left(\frac{h_2 h_3}{h_1} \frac{\partial \mathcal{A}}{\partial u_1} \right) + \frac{\partial}{\partial u_2} \left(\frac{h_1 h_3}{h_2} \frac{\partial \mathcal{A}}{\partial u_2} \right) + \frac{\partial}{\partial u_3} \left(\frac{h_1 h_2}{h_3} \frac{\partial \mathcal{A}}{\partial u_3} \right) \right]. \tag{3.9}$$

3.2 Cylindrical Coordinate Systems

3.2.1 Cylindrical Circular Systems

Circular cylinder systems use a circle as the basis curve. These coordinates are then often called polar coordinates and are defined by

$$x = r \cos \vartheta, \quad y = r \sin \vartheta, \quad z = z,$$
$$0 \leq r < \infty, \quad 0 \leq \vartheta < 2\pi, \quad -\infty < z < \infty, \quad (3.10)$$

with x, y, z the coordinates of a 3D Cartesian coordinate system. The isolines are

1. circles for $r = const$,
2. rays for $\vartheta = const$ and
3. planes for $z = const$.

The metric coefficients for the 3D circular cylinder system are defined by (using $u_1 = r, u_2 = \vartheta$, and $u_3 = z$, see (3.7))

$$h_r = 1, \quad h_\vartheta = r, \quad \text{and} \quad h_z = 1. \quad (3.11)$$

The gradient operator for a scalar field $\mathcal{A}$ is given in cylindrical circular coordinates by

$$\{\nabla \mathcal{A}\}_r = \frac{\partial \mathcal{A}}{\partial r}, \quad \{\nabla \mathcal{A}\}_\vartheta = \frac{1}{\rho} \frac{\partial \mathcal{A}}{\partial \vartheta}, \quad \{\nabla \mathcal{A}\}_z = \frac{\partial \mathcal{A}}{\partial z}. \quad (3.12)$$

The Laplace operator for a scalar field $\mathcal{A}$ is given in cylindrical circular coordinates by

$$\Delta \mathcal{A} = \frac{\partial^2 \mathcal{A}}{\partial r^2} + \frac{1}{r} \frac{\partial \mathcal{A}}{\partial r} + \frac{1}{r^2} \frac{\partial^2 \mathcal{A}}{\partial \vartheta^2} + \frac{\partial^2 \mathcal{A}}{\partial z^2}. \quad (3.13)$$

The problems presented in this work are often independent of the 3rd coordinate z. Then the problem can be reduced to a 2D problem. For the polar 2D case the complex notation (see also Fig. 3.1)

$$\mathbf{z} = x + iy = re^{i\vartheta} \quad (3.14)$$

is commonly used due to its simple notation. An analytic function $\mathbf{w} = \mathbf{f}(\mathbf{z}) = u(x, y) + iv(x, y)$ of the complex variable $\mathbf{z}$ must fulfill the Cauchy-Riemann conditions (see e.g. [3])

$$\frac{\partial u}{\partial x} = \frac{\partial v}{\partial y} \quad \text{and} \quad \frac{\partial u}{\partial y} = -\frac{\partial v}{\partial x}. \quad (3.15)$$

Then the power series for a variable $\mathbf{w}$ is given by

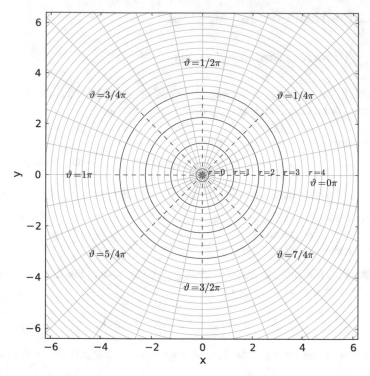

Fig. 3.1 The polar coordinates. Isocurves are given for $r = const$ *(circles)* and for $\vartheta = const$ *(rays)*

$$\mathbf{w} = \mathbf{f}(\mathbf{z}) = u(x, y) + \mathrm{i}v(x, y) = \sum_{n=1}^{\infty} \mathbf{C_n} \left(\frac{\mathbf{z}}{R_{\mathrm{Ref}}} \right)^{n-1}, \qquad (3.16)$$

with the coefficients $\mathbf{C_n} = B_n + \mathrm{i}A_n$. The factor R_{Ref} is typically used as scaling parameter to obtain a polynomial of a dimensionless variable $\mathbf{z}/R_{\mathrm{Ref}}$. R_{Ref} is called the reference radius.

3.2.2   Cylindrical Elliptical Systems

Elliptical shapes are common in accelerator applications, since, e.g., the beam pipe of many accelerators is elliptical and thus elliptic coordinates reflect the geometry of the available aperture. So the reference curve is an ellipse

$$\frac{x^2}{a^2} + \frac{y^2}{b^2} = 1 \qquad (3.17)$$

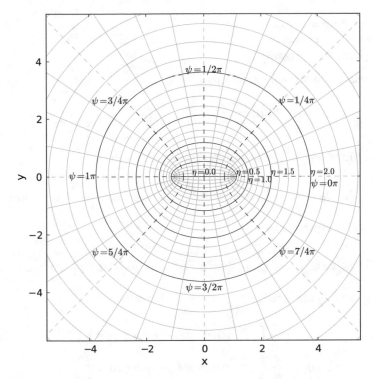

Fig. 3.2 Elliptical coordinates illustrated. Isolines are drawn for various values of η and ψ. The ellipse $\eta = 0$ degenerates to a line. The focal length was set to $e = 1$

with a and b the half axes of the ellipse. Elliptical coordinates are to be used (see Fig. 3.2, see e.g.[1]):

$$x = e \cosh \eta \cos \psi \quad y = e \sinh \eta \sin \psi \quad z = z$$
$$0 \leq \eta < \infty \quad 0 \leq \psi < 2\pi \quad -\infty < z < \infty \quad (3.18)$$

with e the eccentricity of the ellipse which relates to the half axes by

$$b = e \sinh \eta_0, \quad a = e \cosh \eta_0 \quad (3.19)$$

with η_0 the reference ellipse (similar to R_{Ref} for the polar coordinates). It is obtained by

$$\eta_0 = \tanh^{-1}\left(\frac{b}{a}\right). \quad (3.20)$$

The isolines are now

1. ellipses for $\eta = const$,
2. hyperbolas for $\psi = const$,
3. and planes for $z = $ const.

One defines the numerical eccentricity $\hat{\varepsilon}$ by

$$\hat{\varepsilon} = \frac{e}{a} = \frac{\sqrt{a^2 - b^2}}{a} = \sqrt{1 - \tanh^2(\eta_0)}, \qquad e = a\hat{\varepsilon}, \qquad (3.21)$$

with e the focal length. The metric coefficients are given by

$$h_t = h_\eta = h_\psi = e\sqrt{\cosh^2 \eta - \cos^2 \psi} \quad \text{and} \quad h_z = 1. \qquad (3.22)$$

h_t can be also expressed by

$$h_t = e \sqrt{\cosh^2 \eta \, \sin^2 \psi + \sinh^2 \eta \, \cos^2 \psi} = e \sqrt{\sinh^2 \eta + \sin^2 \psi}. \qquad (3.23)$$

The tangents $\vec{q}_{\eta,\psi}$ are obtained by inserting (3.18) into (3.6) to

$$\vec{q}_\eta = \frac{e}{h_t} \left(\sinh \eta \cos \psi \, \vec{i}_x + \cosh \eta \sin \psi \, \vec{i}_y \right), \qquad (3.24)$$

$$\vec{q}_\psi = \frac{e}{h_t} \left(-\cosh \eta \sin \psi \, \vec{i}_x + \sinh \eta \cos \psi \, \vec{i}_y \right),$$

with $\vec{i}_x$ and $\vec{i}_y$ the unit vector in x any y respectively. The gradient is given by

$$\{\nabla\mathcal{A}\}_\eta = \frac{1}{h_t} \frac{\partial \mathcal{A}}{\partial \eta}, \quad \{\nabla\mathcal{A}\}_\psi = \frac{1}{h_t} \frac{\partial \mathcal{A}}{\partial \psi}, \quad \{\nabla\mathcal{A}\}_z = \frac{\partial \mathcal{A}}{\partial z}. \qquad (3.25)$$

Contrary to the polar coordinates (3.11) the metric coefficients of this system, h_η and h_ψ, both depend on η and ψ. The Laplace Operator of a scalar field $\mathcal{A}$ is given by

$$\Delta\mathcal{A} = \frac{1}{e^2 \left(\cosh^2 \eta - \cos^2 \psi \right)} \left[\frac{\partial^2 \mathcal{A}}{\partial \eta^2} + \frac{\partial^2 \mathcal{A}}{\partial \psi^2} \right] + \frac{\partial^2 \mathcal{A}}{\partial z^2}. \qquad (3.26)$$

If the problem to treat is invariant in z, the coordinates and their relation can be described by complex coordinates

$$\mathbf{z} = x + \mathrm{i}y = e \cosh \mathbf{w} = e \cosh (\eta + \mathrm{i}\psi), \qquad (3.27)$$

and the inverse by

$$\mathbf{w} = \cosh^{-1} \left(\frac{1}{e} \mathbf{z} \right) = \cosh^{-1} \left(\frac{1}{e} (x + \mathrm{i}y) \right). \qquad (3.28)$$

Not every available numerical implementation of the complex inverse of the cosine hyperbolic function will produce accurate results; in particular for $\eta = 0$. For example the implementation found in the GNU Scientific Library [4] handles this function in such a way that it can be used to transform elliptical coordinates to Cartesian ones. A full description on its implementation is given in [5, 6]. The numpy package [7], despite its clever implementation and scientific reputation, is implemented based on the functions given in [8], and thus fails to give the correct values. The numerical problems of these equations are described by Hull et al. in the aforementioned papers.

3.3 Toroidal Coordinate Systems

The systems considered up to this point are cylinders and are typically used to solve problems were the dependence on the coordinate z is not given or can be neglected. Similar problems have to be solved for systems were the third coordinate is curved like a circle. An example is the field around the ideal orbit of a particle traversing a dipole.

3.3.1 Global Toroidal Coordinates

Global toroidal coordinates are not further used within this treatise but only given here for completeness. The transformation to Cartesian coordinates is given by (rotating around the y axis, see [1])

$$
\begin{aligned}
x &= \frac{a \sinh \eta \cos \psi}{\cosh \eta - \cos \theta}, \\
y &= \frac{a \sinh \eta \sin \psi}{\cosh \eta - \cos \theta}, \\
z &= \frac{a \sinh \theta}{\cosh \eta - \cos \theta}.
\end{aligned}
\tag{3.29}
$$

3.3.2 Local Toroidal Coordinates

A torus is constructed revolving a circle along an other circle. If the circle revolved is so small that the axis is not touched, the torus is called a "ring torus". Now coordinates can be introduced with (see also Fig. 3.3, [9–11])

Fig. 3.3 Local toroidal coordinate system. Relative local toroidal coordinates ρ, ϑ, φ are given next to the local Cartesian coordinate system x, y, z, and the global Cartesian coordinate system X, Y and Z. Graphic courtesy of B. Seiwald [12]

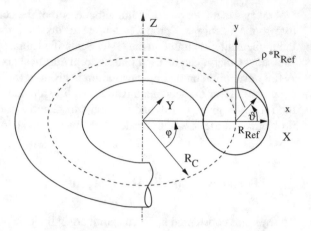

$$X = R_C\, h\, \cos\varphi, \tag{3.30}$$

$$Y = R_C\, h\, \sin\varphi, \tag{3.31}$$

$$Z = \rho\, R_{\text{Ref}}\sin\vartheta, \tag{3.32}$$

$$h = 1 + \varepsilon\rho\cos\vartheta. \tag{3.33}$$

X, Y, Z form a global Cartesian coordinate system, with X and Y defining the plane of the major circle, and with R_C the major radius. R_{Ref} is the radius of the minor circle. The dimensionless constant ε, the inverse aspect ratio,

$$\varepsilon := \frac{R_{\text{Ref}}}{R_C} \tag{3.34}$$

reflects the curvature of the torus. ρ is defined by

$$\rho := \frac{r}{R_{\text{Ref}}} \tag{3.35}$$

with r the radius of the minor circle. Thus all variables of this coordinate system are dimensionless. $h R_C = R_C + x$ gives the distance to the centre in the global Cartesian coordinate system. This coordinate system will be defective for $h = 0$. This is the case for

$$h := 0 \quad \rightarrow \quad R_{\text{Ref}} = R_C, \quad \vartheta = -\pi. \tag{3.36}$$

Further a local Cartesian coordinate system x, y, z is defined at the angle φ by

$$x = \rho\, R_{\text{Ref}}\cos\vartheta \quad \text{and} \quad y = \rho\, R_{\text{Ref}}\sin\vartheta. \tag{3.37}$$

The direction of z is selected in such a way that a right hand system is formed. Notice that the y-axis is always parallel to the Z-axis. x will be parallel to X for $\varphi = 0$. In this case z is antiparallel to Y. The metric coefficients are thus given by

$$h_\rho = R_C \, \varepsilon, \qquad h_\varphi = R_C \, h, \qquad h_\vartheta = R_C \, \varepsilon \, \rho. \tag{3.38}$$

Apart from a constant factor $1/(R_C^2 \, \varepsilon^2)$ the Laplacian in this coordinate system is given by

$$\Delta \mathcal{A} = \left[\frac{\partial^2 \mathcal{A}}{\partial \rho^2} + \frac{1}{\rho} \frac{\partial \mathcal{A}}{\partial \rho} + \frac{1}{\rho^2} \frac{\partial^2 \mathcal{A}}{\partial \vartheta^2} + \frac{\varepsilon}{h} \left(\cos \vartheta \, \frac{\partial \mathcal{A}}{\partial \rho} - \frac{\sin \vartheta}{\rho} \frac{\partial \mathcal{A}}{\partial \vartheta} \right) \right]. \tag{3.39}$$

Comparison to (3.13) shows that the first 3 terms are equivalent to the Laplace operator for circular cylinder coordinates. The last expression $\varepsilon/h(\ldots)$ depends on the inverse aspect ratio of the curvature of the torus.

This coordinate system allows a more straightforward field description as the global toroidal coordinates. It is only applicable for describing the field of magnets with a small ε or $R_{\text{Ref}} \ll R_C$. The beam path within the SIS100 synchrotron falls in that category.

Both Cartesian coordinate systems are typically used for synchrotrons or storage rings; the global Cartesian coordinate system is used for positions in the ring, while the local Cartesian coordinate system is used to describe the field within an accelerator magnet (see e.g. [13–15]). The Cartesian coordinates depicted with X, Y, Z in Fig. 3.3 correspond to the global one while x, y, z corresponds to the local one.

3.3.3 Local Toroidal Elliptical Coordinates

If an ellipse instead of a circle is revolved around the major circle (like pressing down a doughnut), a coordinate system of Local Elliptical Toroidal Coordinates is obtained. Thus the minor circle is replaced by an ellipse.

Its transform to Cartesian coordinates is given by ([16, 17], see also Fig. 3.4)

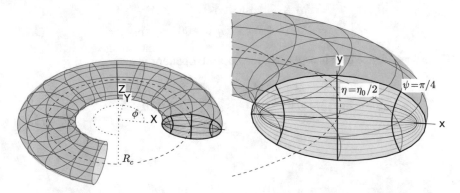

Fig. 3.4 The local elliptical toroidal coordinates. The *left image* gives the total torus while the *right one* shows the local coordinate system

$$X = \left(R_C + \bar{e} \cosh\bar{\eta} \cos\bar{\psi} \right) \cos\phi,$$
$$Y = \left(R_C + \bar{e} \cosh\bar{\eta} \cos\bar{\psi} \right) \sin\phi, \qquad (3.40)$$
$$Z = \bar{e} \sinh\bar{\eta} \sin\bar{\psi},$$

with $\bar{e}$ the eccentricity of the ellipse, $\bar{\eta}$ and $\bar{\psi}$ the coordinates of the ellipse equivalent to η and ψ in (3.18). The eccentricity of the ellipse is equivalent to the one for the elliptical cylinder coordinates (3.21)

$$\bar{e} = \sqrt{a^2 - b^2} \qquad (3.41)$$

with a and b the major and the minor axes of the ellipse. The major radius of the torus is R_C. The boundary of the volume is now an ellipse instead of a circle. Its surface is defined by $\bar{\eta}_0$

$$\tanh\bar{\eta}_0 = \frac{b}{a}. \qquad (3.42)$$

The volume of interest is then limited by :

$$0 \le \bar{\eta} \le \bar{\eta}_0, \qquad -\pi \le \bar{\psi} \le \pi, \qquad \text{and} \qquad -\phi_0 \le \phi \le \phi_0. \qquad (3.43)$$

Equivalent to the ratio of minor to major radius ε one defines

$$\bar{\varepsilon} := \frac{\bar{e}}{R_C} \qquad (3.44)$$

as the ratio of the eccentricity over the major radius. The metric coefficients are then defined by

$$\bar{h}_t = \bar{h}_{\bar{\eta}} = \bar{h}_{\bar{\psi}} = \bar{e}\sqrt{\cosh^2(\bar{\eta}) - \cos^2(\bar{\psi})}, \qquad (3.45)$$
$$\bar{h}_\phi = R_C + \bar{e} \cosh\bar{\eta} \cos\bar{\psi} \qquad (3.46)$$
$$= R_C \left(1 + \bar{\varepsilon} \cosh\bar{\eta} \cos\bar{\psi} \right) = R_C \bar{h}. \qquad (3.47)$$

If the dependence in φ is neglected, the Laplace operator is given by

$$\Delta\mathcal{A} = \frac{1}{\cosh(2\bar{\eta}) - \cos(2\bar{\psi})} \left[\frac{\partial^2\mathcal{A}}{\partial\bar{\eta}^2} + \frac{\partial^2\mathcal{A}}{\partial\bar{\psi}^2} - \frac{\bar{\varepsilon}}{\bar{h}} \left(\sinh\bar{\eta}\cos\bar{\psi}\frac{\partial\mathcal{A}}{\partial\bar{\eta}} + \cosh\bar{\eta}\sin\bar{\psi}\frac{\partial\mathcal{A}}{\partial\bar{\eta}} \right) \right].$$
$$(3.48)$$

3.4 Frenet-Serret Coordinates

Transverse particle beam physics studies the deviation of a particle from the ideal path. Thus a coordinate system is convenient, which follows the reference orbit of the ideal particle. This is the Frenet-Serret coordinate system (e.g. [18, 19]). It describes a particle path by

$$\vec{r}(s) = \vec{r}_0(s) + \delta_d\,\vec{r}(s), \tag{3.49}$$

with $\vec{r}(s)$ the real particles position at the beam path length s, $\vec{r}_0(s)$ the reference particle position and $\delta_d\,\vec{r}(s)$ the particle offset from the reference position. The unity vector $\vec{i}_x(s)$ is perpendicular to the trajectory, and the unity vector $\vec{i}_s(s) = \frac{d\vec{r}_0(s)}{ds}$ is tangential to the trajectory. $\vec{i}_y(s) = \vec{i}_s(s) \times \vec{i}_x(s)$ form an orthogonal right hand basis of this coordinate system (see Fig. 3.5). The change in the vectors is defined within the curvature

$$\frac{d\vec{i}_x(s)}{ds} = k_x\vec{i}_s(s) \quad \text{and} \quad \frac{d\vec{i}_y(s)}{ds} = k_y\vec{i}_s(s), \tag{3.50}$$

with k_x and k_y the curvatures in the xs and ys planes. Any particle position is now given by

$$\vec{r}(x, y, s) = \vec{r}_0(s) + x(s)\,\vec{i}_x(s) + y(s)\,\vec{i}_y(s) \tag{3.51}$$

The Laplace operator in this coordinate frame is not used within this document. The interested reader is referred to [20].

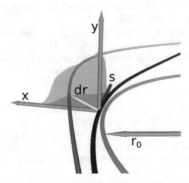

Fig. 3.5 The Frenet-Serret coordinate system. The vector r_0 points to the location where the ideal particle would be. An orthogonal basis x,y,s is moving together with this particle. At any time the vector dr gives the offset between the ideal particle and the real one. The different surfaces (*blue* and *grey*) indicate the planes xy, ys, xs of the coordinate basis

3.5   Summary

The following text is devoted to describe the field within a synchrotron, in particular
the SIS100, consistently. Given that the vacuum chamber is elliptical and the beam
curved within the dipoles, the Local Elliptical Toroidal coordinates are the most
adequate. But already simpler solutions (see Chap. 8 and in particular Sect. 8.2.4) are
accuracte enough for describing the magnetic field of the SIS100. Other machines,
proposed, e.g. the SIS300 or the upgrade of the SPS will use curved dipoles with a
round aperture, where Local Toroidal Coordinates are appropriate.

The metric coefficients allow calculating the Laplace operator for arbitrary curvilinear coordinates. These were presented for the different geometries as the field
within the beam pipe of the magnet is described by solutions of the potential equation (see Sect. 2.1.2)

$$\Delta \Phi = 0. \tag{3.52}$$

The metric coefficients already indicate the complexity increase for the more
advanced systems. Only the metric coefficients of the cylindrical circular coordinate
system are constant or depend only on one variable (see (3.11)), and in particular
only h_ϑ. For the elliptical ones both h_η and h_ψ depend on η and ψ. In case of the
local toroidal ones, the metric coefficients get even more complex. Similar reasoning
applies to the Laplace operator. Thus special methods are required for solving the
Laplace operator for the toroidal coordinates.

References

1. P. Moon, *Field Theory Handbook: Including Coordinate Systems, Differential Equations and Their Solutions* (Springer, Berlin, 1988)
2. W. Papousek, *Vektor Tensor Rechnung*, vol. 2 (Skriptenreferat der Österreichischen Hochschülerschaft, Technische Hochschule in Graz, 1975)
3. E. Kreyszig, *Advanced Engineering Mathematics* (John Wiley and Sons Inc, New York, 1962)
4. M. Galassi, J. Davies, J. Theiler, B. Gough, G. Jungmann, P. Alken, M. Booth, F. Rossi, *GNU scientific library reference manual* (Network Theory Limited, Jan. 2009)
5. T.E. Hull, T.F. Fairgrieve, P.T.P. Tang, Implementing the complex arcsine and arccosine functions using exception handling. ACM Trans. Math. Softw. **23**(3), 299–335 (1997)
6. T.E. Hull, T.F. Fairgrieve, P.T.P. Tang, Implementing complex elementary functions using exception handling. ACM Trans. Math. Softw. **20**(2), 215–244 (1994). June
7. Numerical Python. Hypertext Document at http://www.pfdubois.com/numpy/
8. M. Abramowitz, I. Stegun, *Handbook of Mathematical Functions* (Dover Publication Inc., New York, 1964)
9. D.C. Chang, E.F. Kuester, *Electromagnetic Waves and Curved Structures, Volume 2. IEE Electromagnetic Wave Series* (P. Peregrinus on behalf of the Institution of Electrical Engineers, California, 1977)
10. W.D. D'haeseleer, W.N.G. Hitchon, J.D. Callen, J.-L. Shohet, *Flux Coordinates and Magnetic Field Structure* (Springer, Berlin, 1990)
11. P. Schnizer, B. Schnizer, P. Akishin, E. Fischer, Plane elliptic or toroidal multipole expansions for static fields. Applications within the gap of straight and curved accelerator magnets. Int. J. Comput. Math. Electr. Eng. (COMPEL) **28**(4), 1044–1058 (2009)

12. B. Seiwald, *On Magnetic Fields and MHD Equilibria in Stellarators*. PhD thesis, Technische Universität Graz, Sept. 2007
13. R.A. Beth, Complex representation and computation of two-dimensional magnetic fields. Appl. Phys. **37**, 2568–71 (1966)
14. A.K. Jain, Basic theory of magnets, *CAS Magnetic Measurement and Alignment* (Brookhaven National Laboratory, USA, 1998), pp. 1–21
15. P. Schnizer, Measuring system qualification for LHC arc quadrupole magnets. PhD thesis, TU Graz, 2002
16. P. Schnizer, B. Schnizer, P. Akishin, E. Fischer, Toroidal circular and elliptic multipole expansions within the gap of curved accelerator magnets, *14th International IGTE Symposium, Graz, Institut für Grundlagen und Theorie der Elektrotechnik* (Technische Universität Graz, Austria, 2010)
17. P. Schnizer, B. Schnizer, P. Akishin, A. Mierau, E. Fischer, SIS100 dipole magnet optimisation and local toroidal multipoles. IEEE T. Appl. Supercon. **22**(3), 4001505–4001505 (2012). June
18. H. Wiedemann, *Particle Accelerator Physics* (Springer, Berlin, 2007)
19. K. Wille, *Physik der Teilchenbeschleuniger und Synchrotronstrahlungsquellen* (Teubner, Wiesbaden, 1992)
20. A. C. Kabel, Maxwell-Lorentz equations in general Frenet-Serret coordinates, in *Proceedings of the 2003 Particle Accelerator Conference*, (2003) pp. 2252–2254

Chapter 4
Field Descriptions

Accelerators require beam guiding elements with sufficient field quality. Field deterioration will generate resonances on the beam (for circular machines like synchrotrons or storage rings) or distort the imaging quality as required for analysis magnets (e.g. used for fragment separation).

Typically accelerator beam dynamic books are based on the paraxial approximation and model the guiding elements similar to thin lenses in geometrical optics. The magnetic field is then described by the Taylor series

$$\mathcal{B}(s) = B_0 + K \frac{\partial B(s)}{\partial x} + L \frac{\partial^2 B(s)}{\partial x^2} + \dots, \tag{4.1}$$

with B_0 the constant term and K, L the coefficients of the first higher order contribution, which are then to be determined by some measurements. Measuring derivatives directly is challenging, especially for the accuracy needed in accelerator magnets [1]. The coefficients K, L and so forth are typically obtained using circular cylindrical coordinates (see Sect. 3.2.1) using (2.55), after the field has been integrated over the longitudinal coordinate. This means, however, that neither the curvature of the particle path in the field nor its axis offset variation will be considered.

When a new accelerator was to be built, the magnet aperture was typically designed large enough, so that the magnet's field and, thus, the transverse beam dynamics were safely meeting the projects demands. Large scale projects tend to reduce the magnet apertures as much as possible as it defines the total machine size and costs. Therefore thorough investigations were made for these large machines to estimate the expected machine performance and to prove that a multi-million project would deliver the project's objectives. These machines had typically a circular aperture and were large; thus 2D circular harmonics were used.

The standard literature on beam dynamics (e.g. Wiedemann [2]) develops the magnetic field in a Taylor series (4.1) then just substitutes the coefficients of (4.1) by 2D circular harmonics, stating that these can be used without loss of precision.

© Springer International Publishing AG 2017
P. Schnizer, *Advanced Multipoles for Accelerator Magnets*, Springer Tracts in Modern Physics 277, DOI 10.1007/978-3-319-65666-3_4

Both approaches fail to explain why the coefficients of 2D circular harmonics can be substituted for the coefficients of the Taylor Series. The field descriptions given here and their interpretation will support this choice.

The field description based on circular multipoles was given in Sect. 2.1.2. The content of this chapter concentrates on advanced multipoles as published in [5, 7–9, 11, 15, 20, 21, 25].

4.1 Basis: Cylindrical Circular Multipoles

Circular multipoles are the standard approach used to describe the magnetic field of accelerator magnets. Its complex formulation and its solution, given in Sects. 3.2.1 and 2.1.2, are standard analysis tools. The following derivations will aid the understanding of the toroidal circular multipoles. It was proposed to use complex cylindrical circular multipoles for describing the field of accelerator magnets [12]. The complex potential $\Phi^C(\mathbf{z})$ is given by

$$\Phi^C(\mathbf{z}) = -\sum_{n=1}^{\infty} \frac{1}{n} [B_n + iA_n] \left(\frac{\mathbf{z}}{R_{\text{Ref}}}\right)^n \tag{4.2}$$

with $\mathbf{z} = x + iy$. The field basis functions are then obtained using the gradient

$$\mathbf{B} = -R_{\text{Ref}} \, d\Phi^C(\mathbf{z})/d\mathbf{z}. \tag{4.3}$$

The real potential $\Phi_r(x, y)$ is given by

$$\Phi_r(x, y) = \text{Im}\left[\Phi^C(\mathbf{z})\right]. \tag{4.4}$$

The basis terms are obtained sorting them for B_n and A_n. The B_n are called the "normal" multipoles, while the A_n are called the "skew" multipoles. Its common complex representation is $\mathbf{C_n} = B_n + iA_n$. Then the basis functions for the field are obtained using the gradient

$$B_x(x, y) = -R_{\text{Ref}} \frac{\partial \Phi_r(x, y)}{\partial x} \quad \text{and} \quad B_y(x, y) = -R_{\text{Ref}} \frac{\partial \Phi_r(x, y)}{\partial y}. \tag{4.5}$$

The lowest order terms are listed in Table 4.1. The field generated by these is given in Figs. 4.1 and 4.2. The coefficients are typically obtained from the field on a circle with radius R and calculating the harmonics by

$$\mathbf{C_n} = \frac{1}{2\pi} \int_{-\pi}^{\pi} \mathbf{B}(R, \vartheta) \, e^{-i(n-1)\vartheta} \, d\vartheta. \tag{4.6}$$

Similarly the harmonics can be obtained from the scalar potential $\Phi(R, \vartheta)$ by

Table 4.1 Cylindrical circular multipoles, lower order terms of the potential and the basis functions

	$\Phi_r(x,y)$	$B_x(x,y)$	$B_y(x,y)$
Normal			
1	$-\dfrac{y}{R_{Ref}}$	0	1
2	$-\dfrac{xy}{R_{Ref}^2}$	$\dfrac{y}{R_{Ref}}$	$\dfrac{x}{R_{Ref}}$
3	$\dfrac{1}{3}\dfrac{y(-3x^2+y^2)}{R_{Ref}^3}$	$2\dfrac{xy}{R_{Ref}^2}$	$\dfrac{x^2-y^2}{R_{Ref}^2}$
4	$\dfrac{xy(-x^2+y^2)}{R_{Ref}^4}$	$\dfrac{y(3x^2-y^2)}{R_{Ref}^3}$	$\dfrac{x(x^2-3y^2)}{R_{Ref}^3}$
5	$\dfrac{1}{5}\dfrac{y(-5x^4+10x^2y^2-y^4)}{R_{Ref}^5}$	$4\dfrac{xy(x^2-y^2)}{R_{Ref}^4}$	$\dfrac{x^4-6x^2y^2+y^4}{R_{Ref}^4}$
Skew			
1	$-\dfrac{x}{R_{Ref}}$	1	0
2	$\dfrac{1}{2}\dfrac{-x^2+y^2}{R_{Ref}^2}$	$\dfrac{x}{R_{Ref}}$	$-\dfrac{y}{R_{Ref}}$
3	$\dfrac{1}{3}\dfrac{x(-x^2+3y^2)}{R_{Ref}^3}$	$\dfrac{x^2-y^2}{R_{Ref}^2}$	$-2\dfrac{xy}{R_{Ref}^2}$
4	$\dfrac{1}{4}\dfrac{-x^4+6x^2y^2-y^4}{R_{Ref}^4}$	$\dfrac{x(x^2-3y^2)}{R_{Ref}^3}$	$\dfrac{y(-3x^2+y^2)}{R_{Ref}^3}$
5	$\dfrac{1}{5}\dfrac{x(-x^4+10x^2y^2-5y^4)}{R_{Ref}^5}$	$\dfrac{x^4-6x^2y^2+y^4}{R_{Ref}^4}$	$4\dfrac{xy(-x^2+y^2)}{R_{Ref}^4}$

$$\mathbf{C_n} = \frac{1}{n}\frac{1}{2\pi}\int_{-\pi}^{\pi}\Phi(R,\vartheta)\,e^{-in\vartheta}\,d\vartheta. \tag{4.7}$$

These two methods are commonly used for obtaining the coefficients, where both use the circle as reference curve. The accuracy of the field, described by (2.42) or (4.8), needs to be accurate to typically 100 ppm or better for accelerator applications, therefore the coefficients $\mathbf{C_n} = B_n + iA_n$ have to be known with a similar accuracy.

4.1.1 Conventions

Two conventions are used for describing the field of accelerator magnets with multipoles: The "European" and the "US" one. The European one describes the complex field $\mathbf{B(z)}$ (as already given in (2.42)) by

$$\mathbf{B(z)} = B_y(x+iy) + iB_x(x+iy) = \sum_{n=1}^{N}\mathbf{C_n}\left(\frac{\mathbf{z}}{R_{Ref}}\right)^{(n-1)}, \tag{4.8}$$

with N the highest order multipole. The Taylor series is a infinite one ($N=\infty$). For practical reasons N is typically chosen to not more than 20. This convention is used throughout this document. The US or American conventions starts counting at zero (thus $n' = n - 1$).

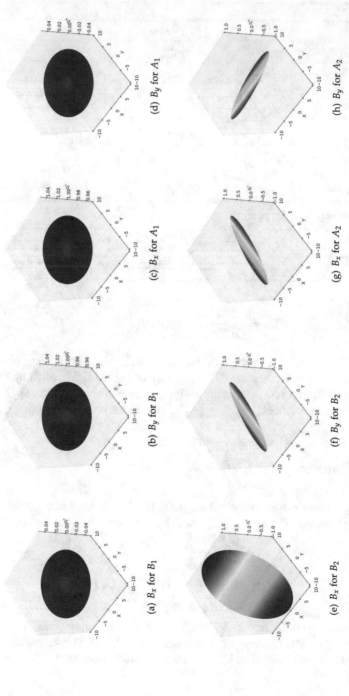

Fig. 4.1 Lowest order terms basis functions for the cylindrical circular multipoles. Here the multipoles are given for $n=1$ (dipole) and $n=2$ (quadrupole). The colour code corresponds to the field B_y or B_x.

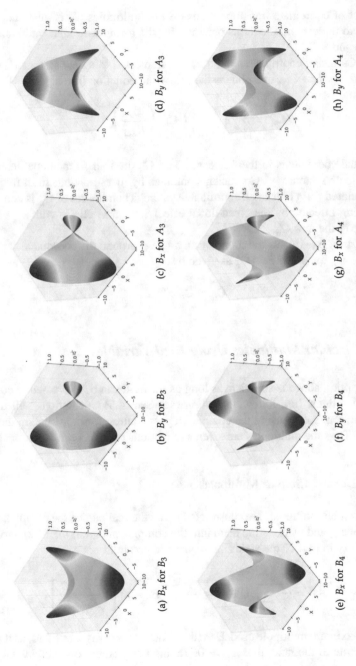

Fig. 4.2 Basis functions for the cylindrical circular multipoles. The multipoles are given for the sextupole (n=3) and the octupole (n=4). The colour code corresponds to the field B_y or B_x.

(a) B_x for B_3

(b) B_y for B_3

(c) B_x for A_3

(d) B_y for A_3

(e) B_x for B_4

(f) B_y for B_4

(g) B_x for A_4

(h) B_y for A_4

4.1.2 Effect of Transformations

The effect of coordinate transformations on the multipoles $\mathbf{C_n}$ is given here, as it will help to understand the terms obtained for the toroidal circular multipoles (see Sects. 4.3 and 8.2).

If the Cartesian coordinate system x, y is translated to a new system x', y' by a distance $\mathbf{d_z} = d_x + id_y$, one obtains the following equation (see e.g. [3, 4]) for the translation

$$\mathbf{C'_n} = \sum_{k=n}^{\infty} \binom{k-1}{k-n} \mathbf{C_k} \left(\frac{\mathbf{d_z}}{R_{\text{Ref}}}\right)^{k-n} , \qquad (4.9)$$

with $\mathbf{C_k}$ the coefficients in the frame x, y and $\mathbf{C'_n}$ the transformed coefficients for the coordinate system $x'y'$. Coefficients obtained by measurements must frequently be recalculated for a translated coordinate system. Then the set of $\mathbf{C'_n}$ is commonly called being affected by the "feed-down effect", because all $\mathbf{C_k}$ with $k > n$ will contribute to the new $\mathbf{C'_n}$.

Similarly if one rotates the frame by an angle α around the coordinate centre the multipoles $\mathbf{C_n}$ transform into $\mathbf{C'_n}$ as given by

$$\mathbf{C'_n} = \mathbf{C_n}\, e^{in\alpha} . \qquad (4.10)$$

4.1.3 Circular Multipoles Using Real Variables

The description above is sufficient as long as the field can be expressed in complex analytic functions. In a following section (see Sect. 4.3) calculations will have to be made in real space, using the complex representation for constructing the basis functions. Thus the multipoles are given as functions of real coordinates.

4.1.3.1 Normal Circular Multipoles

Normal circular multipoles are constructed using the real part of the complex multipole $\mathbf{C_m} = B_m$ and by inserting them into the complex potential Φ^C (4.2), replacing $\mathbf{z}$ with $x + iy$ in (4.4). This yields the scalar real potential

$$\Phi^{Cn} = -\sum_{n=1}^{N} B_n \Phi_n^{Cn} = -\sum_{n=1}^{N} B_n \frac{1}{n} \text{Im}\left[\left(\frac{\mathbf{z}}{R_{\text{Ref}}}\right)^n\right]. \qquad (4.11)$$

N is the maximum multipole used. Here the constant part was set to zero, as it would not contribute to the field, thus $B_0 = 0$. Taking the gradient one obtains the basis functions for the field

$$(B_x^{Cn}, B_y^{Cn}) = -R_{\text{Ref}} \, \nabla \Phi^{Cn}(x, y) = -R_{\text{Ref}} \left(\frac{\partial}{\partial x}, \frac{\partial}{\partial y} \right) \Phi^{Cn}(x, y). \qquad (4.12)$$

These are identical if only B_n is used in (4.8) and the real part is used for B_y and the imaginary part for B_x. The scalar potential for the normal dipole is obtained for $n = 1$, the one for the quadrupole for $n = 2$ and so on.

4.1.3.2 Skew Circular Multipoles

Similarly the potential for the different skew harmonics can be given using $C_n = iA_n$ in (4.4). Then the potential for the different skew harmonics $\Phi^{Cs}(x, y)$ is obtained by

$$\Phi^{Cs}(x, y) = - \sum_{n=1}^{N} A_n \Phi_n^{Cs} = - \sum_{n=1}^{N} A_n \frac{1}{n} \text{Re} \left[\left(\frac{z}{R_{\text{Ref}}} \right)^n \right] = - \sum_{n=1}^{N} A_n \frac{1}{n} \text{Im} \left[i \left(\frac{z}{R_{\text{Ref}}} \right)^n \right].$$
$$(4.13)$$

The constant term is set to zero, thus $A_0 = 0$. Then the field is given by B_x^{Cs}, B_y^{Cs}

$$(B_x^{Cs}, B_y^{Cs}) = -R_{\text{Ref}} \, \nabla \Phi^{Cs}(x, y) = -R_{\text{Ref}} \left(\frac{\partial}{\partial x}, \frac{\partial}{\partial y} \right) \Phi^{Cs}(x, y). \qquad (4.14)$$

The obtained functions are identical if iA_n is inserted in (4.8) and the real part is used for B_y and the imaginary part for B_x.

4.2 Cylindrical Elliptical Multipoles

The aperture of the vacuum chamber of SIS100 is elliptical. While circular multipoles allow describing the field within this aperture, the elliptical multipoles are the more natural ones. The work summarised here was presented in [5–9] (first reports [10, 11]). Related work of other authors can be found in [12–14]. The deduction here is based on [6, 15].

If the dependence on z can be neglected (3.26) reduces to

$$\Delta \Phi^e = \frac{1}{e^2 \left(\cosh^2 \eta - \cos^2 \psi \right)} \left[\frac{\partial^2}{\partial \eta^2} + \frac{\partial^2}{\partial \psi^2} \right] \Phi^e. \qquad (4.15)$$

The ansatz $\Phi^e(\eta, \psi) = H(\eta) \, \Psi(\psi)$ yields the two ordinary differential equations

$$\frac{d^2 H}{d\eta^2} - \gamma H = 0 \quad \text{and} \quad \frac{\partial^2 \Psi}{\partial \psi^2} + \gamma \Psi = 0.$$

γ is the separation constant. These equations can be solved by using the ansatz for all $m \neq 0$

$$\Psi_m(\psi) = e^{im\psi}, \qquad H_m(\eta) = a_m \cosh(m\eta) + b_m \sinh(m\eta); \qquad m = \pm1, \pm2, \pm3, \ldots . \tag{4.16}$$

m equals zero yields

$$\Psi_0 = 1; \quad H_0(\eta) = a_0 + b_0\, \eta. \tag{4.17}$$

So the total solution is given by

$$\Psi_g(\eta, \psi) = a_0 + b_0\eta + \sum_{m=-\infty,\, m\neq0}^{m=\infty} [a_m \sinh(|m|\eta) + b_m \cosh(|m|\eta)]\, e^{im\psi}. \tag{4.18}$$

Here the term b_0 must be 0 as it will not be a harmonic solution [6, 16]. The sum does not contain the term $m = 0$, thus the constant field is given by the term a_0. If the potential is defined on the boundary, the coefficients a_m and b_m are obtained by (4.18)

$$b_m \cosh(|m|\eta_0) + a_m \sinh(|m|\eta_0) = \frac{1}{2\pi} \int_{-\pi}^{\pi} \Psi(\eta_0, \psi)\, e^{-im\psi}\, d\psi. \tag{4.19}$$

The $\cosh(|m|\eta_0)$ and $\sinh(|m|\eta_0)$ are used to scale the coefficients b_m and a_m. This equation will not yield sufficient conditions. The right hand side yields only a single value while on the left hand side two coefficients are to be determined, thus no unique solution can be obtained. But as shown in text following (4.22) the terms

$$1; \quad \cos(m\psi)\cosh(m\eta), \quad \sin(m\psi)\sinh(m\eta); \qquad m = 1, 2, 3, \ldots \tag{4.20}$$

give a sufficient general ansatz to describe all possible solutions. So the required solution is given by

$$\Psi(\eta, \psi)/\Psi_0 = \frac{a_0}{2} + \sum_{m=1}^{\infty} \left[a_m \frac{\cosh(m\eta)}{\cosh(m\eta_0)} \cos(m\psi) + b_m \frac{\sinh(m\eta)}{\sinh(m\eta_0)} \sin(m\psi) \right], \tag{4.21}$$

and the coefficients a_m and b_m are obtained by

$$a_m = \frac{1}{\pi} \int_{-\pi}^{\pi} \frac{\Psi(\eta_0, \psi)}{\Psi_0} \cos(m\psi)\, d\psi, \qquad b_m = \frac{1}{\pi} \int_{-\pi}^{\pi} \frac{\Psi(\eta_0, \psi)}{\Psi_0} \sin(m\psi)\, d\psi. \tag{4.22}$$

These multipoles are a complete set as are the circular ones [11, 17]. For a first indication one can use the definition of the elliptical coordinates (3.18) and inserts them into the description of the magnetic field using circular multipoles (3.16) (here with $R_{\mathrm{Ref}} = 1$). Then one sorts the result in the different trigonometric and hyperbolic terms which yields for example for $n = 4$

$$4\,\mathrm{Re}((x+iy)^3) = 4x^3 - 12xy^2 = e^3\,[3\cosh(\eta)\cos(\psi) + \cosh(3\eta)\cos(3\psi)],$$
$$4\,\mathrm{Im}((x+iy)^3) = 12x^2y - 4y^3 = e^3\,[3\sinh(\eta)\sin(\psi) + \sinh(3\eta)\sin(3\psi)].$$
$$(4.23)$$

The first line describes the normal octupole while the second one describes the skew one. Thus it can be seen that only terms as given in (4.20) are left over. In this way one may find an infinite matrix which transforms the elliptic multipoles into circular ones and vice versa. All the matrix elements are derived below (see Sect. 4.2.2). A second proof of the completeness of the elliptic multipoles is given indirectly in [18] pp. 1202 f. These authors derive the Green's function of the potential equation in cylindrical elliptical coordinates. There they also derive an eigen function expansion which exactly contains all the functions listed in (4.20). Such expansions always use a complete system of eigenfunctions.

4.2.1 Complex Elliptical Multipoles

Complex notation simplifies the representation of the field. Replacing sin and cos and using the addition theorem for the complex cosh one can write Eq. (4.20) by

$$\mathbf{B} = B_y(\eta + i\psi) + iB_x(\eta + i\psi) = \frac{\mathbf{E_1}}{2} + \sum_{m=2}^{\infty} \mathbf{E_m}\, \frac{\cosh[(m-1)(\eta + i\psi)]}{\cosh[(m-1)\,\eta_0]}$$

$$(4.24)$$

$$= \sum_{m=1}^{\infty} \mathbf{E}_m\, \frac{1}{1+\delta_{m1}}\, \frac{\cosh[(m-1)\mathbf{w}]}{\cosh[(m-1)\eta_0]}$$

$$:= -\,d\,\Xi^{Ce}/d\mathbf{w}\,,$$

$$(4.25)$$

with Ξ^{Ce} the complex potential. The real and imaginary part of the complex coefficient $\mathbf{E_m}$ are defined by $\mathbf{E_m} = E_m^n + iE_m^s$. That the terms are bounded for $0 \le \eta \le \eta_0$ can be shown by

$$\left| \frac{\cosh[(m-1)(\eta+i\psi)]}{\cosh[(m-1)\eta_0]} \right| = \sqrt{\frac{\cos[2(m-1)\psi] + \cosh[2(m-1)\eta]}{1 + \cosh[2(m-1)\eta_0]}} \le \qquad (4.26)$$

$$\le \frac{\cosh[(m-1)\eta]}{\cosh[(m-1)\eta_0]} \sim e^{-(m-1)(\eta_0 - \eta)}. \qquad (4.27)$$

The coefficients $\mathbf{E_m}$ can be calculated by

$$\mathbf{E}_m = \frac{1}{\pi} \int_{-\pi}^{\pi} \mathbf{B_0}\,(\mathbf{z} = e\cosh(\eta_0 + i\psi))\,\cos[(m-1)\,\psi]\,d\psi, \tag{4.28}$$

$$= \frac{1}{2\pi} \int_{-\pi}^{\pi} [\mathbf{B_0}\,(\mathbf{z} = e\cosh(\eta_0 + i\psi)) + \mathbf{B_0}\,(\mathbf{z} = e\cosh(\eta_0 - i\psi))]\,e^{i(m-1)\psi}\,d\psi. \tag{4.29}$$

It can be shown that the expansion (4.24) resembles all required circular terms. For that purpose one inserts the definition of the elliptical coordinates (3.27) into the magnetic field (2.42) (in circular coordinates). This yields then

$$\mathbf{B(z)} = \sum_{n=1}^{\infty} \mathbf{C_n} \left(\frac{x + iy}{R_{\text{Ref}}}\right)^{n-1} = \sum_{m=1}^{\infty} \mathbf{C_m}\,\alpha^{m-1}\,\cosh^{m-1}\mathbf{w} \tag{4.30}$$

with

$$\alpha := \frac{e}{R_{\text{Ref}}}. \tag{4.31}$$

For circular multipoles negative powers are not allowed and any $\cosh^m(x)$ can be expressed by a series of $\cosh(mx)$ [19]. Thus only terms of $\cosh(m\mathbf{w})$ are obtained. The first basis terms are given in Table 4.2. The field generated by the first basis terms is given in Figs. 4.3 and 4.4. The indices m and n in (4.30) are identical. The switch to m is made as in the following the term $\cosh^{m-1}\mathbf{w}$ will be expanded.

Table 4.2 The basis functions of the cylindric elliptical multipoles. These are given for the lower order basis functions for the potential of the normal and skew multipoles $\Xi^{Cen}(\eta, \psi)$, $\Xi^{Ces}(\eta, \psi)$ and for the field B_y and B_x

	$\Xi^{Cen}(\eta, \psi)$, $\Xi^{Ces}(\eta, \psi)$	$B_x(\eta, \psi)$	$B_y(\eta, \psi)$
Normal			
1	$-\frac{\psi}{2}$	0	$\frac{1}{2}$
2	$-\cosh(\eta)\sin(\psi)$	$\sin(\psi)\sinh(\eta)$	$\cos(\psi)\cosh(\eta)$
3	$-\frac{1}{2}\cosh(2\eta)\sin(2\psi)$	$\sin(2\psi)\sinh(2\eta)$	$\cos(2\psi)\cosh(2\eta)$
4	$-\frac{1}{3}\cosh(3\eta)\sin(3\psi)$	$\sin(3\psi)\sinh(3\eta)$	$\cos(3\psi)\cosh(3\eta)$
Skew			
1	$-\frac{\eta}{2}$	$\frac{1}{2}$	0
2	$-\sinh(\eta)\cos(\psi)$	$\cos(\psi)\cosh(\eta)$	$-\sin(\psi)\sinh(\eta)$
3	$-\frac{1}{2}\sinh(2\eta)\cos(2\psi)$	$\cos(2\psi)\cosh(2\eta)$	$-\sin(2\psi)\sinh(2\eta)$
4	$-\frac{1}{3}\sinh(3\eta)\cos(3\psi)$	$\cos(3\psi)\cosh(3\eta)$	$-\sin(3\psi)\sinh(3\eta)$

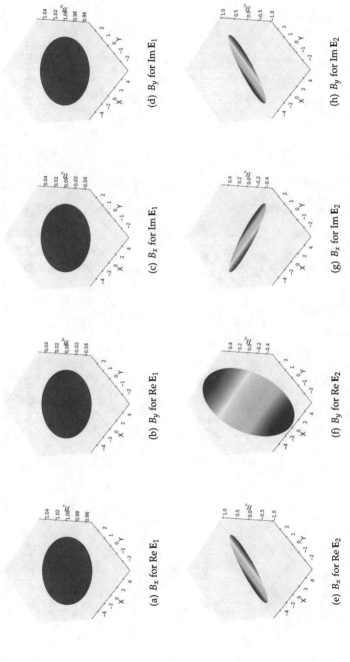

Fig. 4.3 The lowest order basis functions for the cylindrical elliptical multipoles. Here the multipoles are given for $m = 1$ and $m = 2$. The colour code corresponds to the field B_y or B_x.

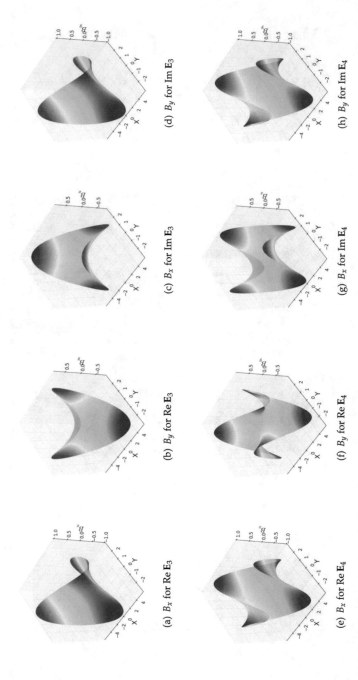

Fig. 4.4 Basis functions for the cylindrical elliptical multipoles. Here the multipoles are given for $m = 3$ and $m = 4$. The colour code corresponds to the field B_y or B_x.

4.2.2 Relations Between Circular and Elliptical Multipoles

The cylindric circular multipoles and the cylindric elliptical multipoles have to represent the same field therefore (4.24) and (4.30) are set equal. The coefficients E_m are obtained using (4.30) along $\eta = 0$.

$$\hat{E}_k := \frac{E_k}{\cosh(k\eta_0)} = \sum_{m=1}^{\infty} C_m \, \alpha^{m-1} \, \bar{s}_{mk}, \tag{4.32}$$

with

$$\bar{s}_{mk} = \delta_{m+k,\,\text{even}} \frac{1}{2^{(m-2)}} \binom{m-1}{(m-k)/2}, \qquad k, m = 1, 2, 3, \dots; \tag{4.33}$$

and

$$\delta_{m+k,\,\text{even}} := [1 + (-1)^{m+k}]/2. \tag{4.34}$$

The integrals of $\cos(m\psi)$ are then obtained by the binomial theorem [13] or found in [19]. The matrix $\bar{s}_{mk}$ is a lower triangular matrix with every second argument being 0. The binomial coefficient vanishes if the lower entry is smaller than 0. The undefined ones are multiplied by 0 due to (4.34).

Normally the elliptical multipoles are obtained from the data on the ellipse. Then the inverse of the matrix $\bar{S}$ is required, which can be obtained numerically. In addition it can be solved by equating the field descriptions using

$$\alpha^m \, C_{m+1} = \sum_{k=1}^{\infty} \hat{E}_k / [(1 + \delta_{k1}) \cosh((k-1)\,\eta_0)] \, t_{km}, \quad k, m = 1, 2, \dots . \tag{4.35}$$

The matrix T is $T = (t_{km}) = \tilde{\bar{S}}^{-1}$. Thus it is the transposed inverse of the matrix $\bar{S} = (\bar{s}_{mk})$.

In a first step one equalises the description for the cylindrical circular (2.42) (or (4.8)) and cylindrical elliptical multipoles (4.24), (4.30):

$$\sum_{n=1}^{\infty} C_n \left(\frac{z}{R_{\text{Ref}}}\right)^{n-1} = \frac{E_1}{2} + \sum_{m=2}^{\infty} \hat{E}_m \cosh[(m-1)w]. \tag{4.36}$$

Then both sides of the equation are divided by z^s and integrated over z

$$\oint_{C_z} \sum_{n=1}^{\infty} \frac{1}{z^s} C_n \left(\frac{z}{R_{\text{Ref}}}\right)^{n-1} dz = \oint_{C_z} \frac{1}{z^s} \left[\frac{E_1}{2} + \sum_{m=2}^{\infty} E_m \cosh[(m-1)w]\right] dz \tag{4.37}$$

along a closed curve C_z surrounding the point $z = 0$. The integral on the left is evaluated by Cauchy's residue theorem. Only the term for $n = s$ is left from the infinite sum. Its value is $2\pi i\ \mathbf{C_s}/R_{\text{Ref}}^{s-1}\ \delta_{sn}$. Now the variable $\mathbf{z}$ is substituted by $e\cosh(\mathbf{w})$ and $d\mathbf{z}$ is substituted by $e\sinh(\mathbf{w})d\mathbf{w}$, so a mapping with the properties $\mathbf{z}=0 \leftrightarrow \mathbf{w_0}=i\pi/2$ is achieved. The curve C_z is mapped on a closed curve around $\mathbf{w_0}$.

The left side of (4.37) gives only one $\mathbf{C_s}$ while on the right side different terms are left over which are collected in the matrix t_{ms}

$$2\pi i\ \frac{\mathbf{C_s}}{R_{\text{Ref}}^{s-1}} = 2\pi i\ \frac{\mathbf{E_1}}{2}\ \frac{t_{1s}}{e^{s-1}} + \sum_{m=2}^{\infty} 2\pi i\ \hat{\mathbf{E}}_m\ \frac{t_{ms}}{e^{s-1}} \tag{4.38}$$

with

$$2\pi i\ t_{ms} = \oint_{C_w} d\mathbf{w}\ \sinh(\mathbf{w})\ \cosh\left[(m-1)\,\mathbf{w}\right]/\cosh^s\mathbf{w}. \tag{4.39}$$

This expression has a pole of order s at $\mathbf{w_0}$. The residue theorem yields

$$t_{ms} = \text{Res}\left(\frac{\sinh\mathbf{w}\ \cosh\left[(m-1)\,\mathbf{w}\right]}{\cosh^s\mathbf{w}},\ \mathbf{w}=\mathbf{w_0}=i\frac{\pi}{2}\right). \tag{4.40}$$

"Res" denotes the residue of the poles at $\mathbf{w}=i\pi/2$. Before evaluating the elements of the matrix t_{ms} it is shown that it is a lower triangular matrix. This is proven by looking at the poles occurring in the function just for $\mathbf{w}=i\pi/2$. For this, $\cosh[(m-1)\mathbf{w}]$ is represented as a sum of powers of $\cosh\mathbf{w}$ [15, 19].

$$\cosh[(m-1)\mathbf{w}] = \sum_{k=0}^{K} c_{m-1-2k}\ \cosh^{m-1-2k}\mathbf{w}. \tag{4.41}$$

The exponents are non-negative integers. The limit K must be chosen accordingly. The values of the coefficients are not needed for the present considerations. $\sinh\mathbf{w} = \sin(\pi/2)$ is just unity. The remaining part of (4.40) is

$$\cosh[(m-1)\mathbf{w}]/\cosh^s\mathbf{w} = \sum_{k=0}^{K} c_{m-1-2k}\ \cosh^{m-1-2k-s}\mathbf{w}. \tag{4.42}$$

This function has poles at $\mathbf{w}=i\pi/2$ only if

$$m-1-2k-s=-1,\quad \text{i.e. for}\quad s\geq m+2k, k=0,1,\ldots,K. \tag{4.43}$$

$k=0$ gives the diagonal of t_{ms}. $k=1,2,\ldots$ gives elements below the diagonal; only each second element is not vanishing. Equation (4.40) can be evaluated using computer algebra systems. This was done for several values of m which allowed deducing the coefficients. It was found that the coefficients t_{ms} can be described by

$$s = \text{even}: \quad t_{ms} = (-1)^{\frac{s-2}{2}} \sin\left((m-1)\frac{\pi}{2}\right) \frac{m-1}{(s-1)!} \prod_{\mu=1}^{\frac{s-2}{2}} \left((m-1)^2 - [2\mu - 1]^2\right),$$

$$(4.44)$$

$$s = \text{odd}: \quad t_{ms} = (-1)^{\frac{s-1}{2}} \cos\left((m-1)\frac{\pi}{2}\right) \frac{1}{(s-1)!} \prod_{\mu=1}^{\frac{s-1}{2}} \left((m-1)^2 - 4(\mu-1)^2\right).$$

$$(4.45)$$

Peña and Franchetti derived a similar formula for the matrix t_{ms} by a recursion method [13] with different expressions, because they used different conventions. Furthermore one could derive them from Chebychev polynoms [15, 25] or other integrals [15].

The final conversion formula is given by

$$\mathbf{C_s} = \sum_{m=1}^{s} \underbrace{\frac{t_{ms}/(1+\delta_{m1})}{\cosh\left[(m-1)\,\eta_0\right]\alpha^s}}_{T_{ms}} \mathbf{E_m}.$$

$$(4.46)$$

The conversion matrix t_{ms} is given for different ellipses in Table 4.3, with the first one corresponding to the beam vacuum chamber size (see Sect. 7.2) and the second to the ellipse covered by measurements (see Sect. 8.1).

The relations, presented above, are given for a field of Cartesian field vectors and elliptical coordinates. The potential $\Xi^{Ce}(\mathbf{w})$ as defined in (4.24) can be obtained by integration in $\mathbf{w}$. It is given by

$$\Xi^{Ce}(\mathbf{w}) = -\frac{\mathbf{E_1}}{2}\mathbf{w} - \sum_{m=2}^{M} \frac{1}{m-1} \mathbf{E_m} \frac{\sinh[(m-1)\,\mathbf{w}]}{\cosh[(m-1)\,\eta_0]}.$$

$$(4.47)$$

M is the highest order multipole taken into account. The integration constant $\Xi^{Ce}(\mathbf{w}=0)=0$ is set to zero. $\Xi^{Ce}(\mathbf{w})$ is not a single valued potential. The Cartesian field vector expansions of the magnetic field can be derived from this potential as given in the next section (see (4.49) and (4.52)). But it is not obtained applying the gradient ∇ in elliptical coordinates on $\Xi^{Ce}(\mathbf{w})$ [15].

4.2.2.1 Normal Multipole Expansions for Cartesian Components

Similar as for the cylindrical circular components the field expansions for the normal components can be obtained using only the $\mathbf{E_m} = E_m^n$ in Eq. (4.24). Then the field components B_y^{Cen} and B_x^{Cen} are given by

Table 4.3 Conversion matrices for two different ellipses. The different factors are given rounded to two digits after the comma. All factors exactly zero are left out

s	m									
	$a = 5.75, b = 3.0, R_{Ref} = 4.0$									
	1	2	3	4	5	6	7	8	9	10
1	0.50		−0.57		0.20		−0.06		0.02	
2		0.70		−0.84		0.45		−0.20		0.08
3			0.76		−1.04		0.74		−0.42	
4				0.74		−1.20		1.06		−0.71
5					0.69		−1.32		1.38	
6						0.64		−1.41		1.70
7							0.58		−1.47	
8								0.53		−1.51
9									0.49	
10										0.45
	$a = 4.5, b = 1.7, R_{Ref} = 4.0$									
1	0.50		−0.75		0.39		−0.18		0.08	
2		0.89		−1.60		1.29		−0.83		0.48
3			1.38		−2.89		3.03		−2.45	
4				1.97		−4.76		6.11		−5.93
5					2.66		−7.45		11.29	
6						3.51		−11.26		19.68
7							4.58		−16.65	
8								5.93		−24.19
9									7.67	
10										9.91

$$B_y^{Cen}(\eta, \psi) = \frac{E_1^n}{2} + \sum_{m=2}^{M} E_m^n \frac{\cosh[(m-1)\,\eta]\,\cos[(m-1)\,\psi]}{\cosh[(m-1)\,\eta_0]},$$

$$(4.48)$$

$$B_x^{Cen}(\eta, \psi) = \sum_{m=2}^{M} E_m^n \frac{\sinh[(m-1)\,\eta]\,\sin[(m-1)\,\psi]}{\cosh[(m-1)\,\eta_0]}$$

This can be calculated from the potential using

$$\left(B_x^{Cen}(\eta, \psi),\, B_y^{Cen}(\eta, \psi)\right) = -\left(\frac{\partial}{\partial \eta}, \frac{\partial}{\partial \psi}\right) \Xi^{Cen}(\eta, \psi). \qquad (4.49)$$

$\Xi^{Cen}(\eta, \psi)$ is the imaginary part of the complex potential $\Xi^e(\mathbf{w})$ (see (4.47)). Thus the potential is given by

$$\Xi^{\text{Cen}}(\eta, \psi) = -\frac{E_1^{\text{n}}}{2}\psi - \sum_{m=2}^{M}\frac{1}{m-1}E_m^{\text{n}}\frac{\cosh[(m-1)\eta]\sin[(m-1)\psi]}{\cosh[(m-1)\eta_0]}. \quad (4.50)$$

4.2.2.2 Skew Multipole Expansions for Cartesian Components

In the same manner as for the normal multipoles, the field expansions can be obtained inserting $\mathbf{E}_m = \mathrm{i}E_m^{\text{s}}$ in (4.24). This yields the field components B_y^{Ces} and B_x^{Ces}

$$B_y^{\text{Ces}}(\eta, \psi) = -\sum_{m=2}^{M}E_m^{\text{s}}\frac{\sinh[(m-1)\eta]\,\sin[(m-1)\psi]}{\cosh[(m-1)\eta_0]},$$

$$(4.51)$$

$$B_x^{\text{Ces}}(\eta, \psi) = \frac{E_1^{\text{s}}}{2} + \sum_{m=2}^{M}E_m^{\text{s}}\frac{\cosh[(m-1)\eta]\,\cos[(m-1)\psi]}{\cosh[(m-1)\eta_0]}.$$

Similarly the potential is defined by

$$\left(B_x^{\text{Ces}}(\eta, \psi), B_y^{\text{Ces}}(\eta, \psi)\right) = -\left(\frac{\partial}{\partial\eta}, \frac{\partial}{\partial\psi}\right)\Xi^{\text{Ces}}(\eta, \psi), \quad (4.52)$$

which can be also obtained using the real part of $\Xi^{Ce}(\mathbf{w})$ (4.47), if $\mathbf{E_m} = \mathrm{i}E_m^{\text{s}}$. This yields

$$\Xi^{Ces}(\eta, \psi) = -\frac{E_1^{\text{s}}}{2}\eta - \sum_{m=2}^{M}\frac{1}{m-1}E_m^{\text{s}}\frac{\sinh[(m-1)\eta]\,\cos[(m-1)\psi]}{\cosh[(m-1)\eta_0]}. \quad (4.53)$$

Ξ^{Ces} is only the imaginary part of $\Xi^e(\mathbf{w})$ (4.47).

4.2.3 Elliptical Multipole Field Expansions for Elliptical Components

The above expressions were given for field components expressed in a Cartesian coordinate system on an elliptical coordinate systems. Here elliptical field components B_η, B_ψ depending on elliptical coordinates are to be derived which describe the same field for the same coefficients $\mathbf{E}$ as for the field components B_y and B_x. So one converts B_y and B_x into B_η, B_ψ using (3.24). For these elliptical field components the potential as a function of elliptical coordinates is given by

$$(B_\eta, B_\psi) = -\frac{1}{h_t}\left(\frac{\partial\Phi^e}{\partial\eta}, \frac{\partial\Phi^e}{\partial\psi}\right). \quad (4.54)$$

These relations are specialised for the normal and skew part below. h_t is the elliptical metric element (3.23).

4.2.3.1   Normal Multipoles

Scaled coefficients $\hat{E}_m^n$ for the normal harmonics are defined by

$$\hat{E}_m^n = \text{Re}[\mathbf{E_m}]/\cosh[(m-1)\,\eta_0] \tag{4.55}$$

to get more concise expressions below. The Cartesian field components (see (4.48)) are translated to elliptical field components B_η^{en} and B_ψ^{en} using (3.24) which yields

$$B_\eta^{en}(\eta,\psi) = \frac{e}{2h_t}\Bigg[\hat{E}_1^n \cosh\eta\,\sin\psi + \hat{E}_2^n \cosh[2\eta]\,\sin[2\psi] + \tag{4.56}$$

$$+ \sum_{m=2}^{M-1} \hat{E}_{m+1}^n\Big(\cosh[(m+1)\,\eta]\sin[(m+1)\,\psi] - \cosh[(m-1)\,\eta]\sin[(m-1)\,\psi]\Big)\Bigg]$$

$$= -\frac{1}{h_t}\frac{\partial\Phi^{en}}{\partial\eta},$$

$$B_\psi^{en}(\eta,\psi) = \frac{e}{2h_t}\Bigg[\hat{E}_1^n \sinh\eta\,\cos\psi + \hat{E}_2^n \sinh[2\eta]\,\cos[2\psi] + \tag{4.57}$$

$$+ \sum_{m=2}^{M-1} \hat{E}_{m+1}^n\Big(\sinh[(m+1)\eta]\cos[(m+1)\psi] - \sinh[(m-1)\eta]\cos[(m-1)\psi]\Big)\Bigg]$$

$$= -\frac{1}{h_t}\frac{\partial\Phi^{en}}{\partial\psi},$$

with e the eccentricity of the ellipse. The potential can be derived by integration and is given by

$$\Phi^{en}(\eta,\psi) = -\frac{e}{2}\Bigg[\hat{E}_1^n \sinh\eta\,\sin\psi + \hat{E}_2^n\,\frac{1}{2}\sinh(2\eta)\,\sin(2\psi) +$$

$$+ \sum_{m=2}^{M-1} \hat{E}_{m+1}^n\left(\frac{\sinh[(m+1)\,\eta]\sin[(m+1)\,\psi]}{m+1} - \frac{\sinh[(m-1)\,\eta]\sin[(m-1)\,\psi]}{m-1}\right)\Bigg].$$

$$\tag{4.58}$$

4.2.3.2 Skew Multipoles

Similar as for the normal harmonics scaled skew coefficients are defined for the skew harmonics by

$$\hat{E}_m^s = \text{Im}[\mathbf{E_m}]/\cosh[(m-1)\,\eta_0] \tag{4.59}$$

The Cartesian field components are obtained from (4.51) which are transformed to B_η^{es} and B_ψ^{es} using (3.24). This gives

$$B_\eta^{es}(\eta, \psi) = \frac{e}{2h_t}\left[\hat{E}_1^s \sinh\eta \, \cos\psi + \hat{E}_2^s \sinh(2\eta) \, \cos(2\psi) + \right. \tag{4.60}$$

$$+ \sum_{m=2}^{M-1} \hat{E}_{m+1}^s \Big(\sinh[(m+1)\,\eta]\cos[(m+1)\,\psi] - \sinh[(m-1)\,\eta]\cos[(m-1)\,\psi] \Big)\Big]$$

$$= -\frac{1}{h_t}\frac{\partial \Phi^{es}}{\partial \eta},$$

$$B_\psi^{es}(\eta, \psi) = -\frac{e}{2h_t}\left[\hat{E}_1^s \cosh\eta \, \sin\psi + \hat{E}_2^s \cosh(2\eta) \, \sin(2\psi) + \right. \tag{4.61}$$

$$+ \sum_{m=2}^{M-1} \hat{E}_{m+1}^s \Big(\cosh[(m+1)\,\eta]\sin[(m+1)\,\psi] - \cosh[(m-1)\,\eta]\sin[(m-1)\,\psi] \Big)\Big]$$

$$= -\frac{1}{h_t}\frac{\partial \Phi^{es}}{\partial \psi}.$$

Integrating the fields the potential can be derived, which is given by

$$\Phi^{es}(\eta, \psi) = -\frac{e}{2}\left[\hat{E}_1^s \cosh\eta \, \cos\psi + \hat{E}_2^s \frac{1}{2}\cosh(2\eta) \, \cos(2\psi) + \right.$$

$$+ \sum_{m=2}^{M-1} \hat{E}_{m+1}^s \left(\frac{\cosh[(m+1)\,\eta]\cos[(m+1)\,\psi]}{m+1} - \frac{\cosh[(m-1)\,\eta]\cos[(m-1)\,\psi]}{m-1} \right)\right].$$

$$\tag{4.62}$$

4.2.4 Complex Potential for Normal and Skew Elliptical Multipoles

The fields B_η^e and B_ψ^e can be expressed more concisely then above defining a complex analytic function $\hat{\mathbf{B}}^e(w)$ by

$$\hat{\mathbf{B}}^e(w) = (h_t\, B_\psi^e + i\, h_t\, B_\eta^e) =$$

$$= \frac{e}{2}\left[\hat{\mathbf{E}}_1 \sinh[w] + \hat{\mathbf{E}}_2 \sinh[2w] + \sum_{m=2}^{M-1} \hat{\mathbf{E}}_{m+1}(\sinh[(m+1)w] - \sinh[(m-1)w]) \right],$$
(4.63)

with $\hat{\mathbf{E}}_m = \mathbf{E_m}/\cosh[(m-1)\,\eta_0]$. This can be rearranged to

$$\hat{\mathbf{B}}^e(w) = e \sinh[w]\left[\frac{\hat{\mathbf{E}}_1}{2} + \sum_{m=2}^{M-1} \hat{\mathbf{E}}_m \cosh[(m-1)w] \right].$$
(4.64)

Integrating $\hat{\mathbf{B}}^e(w)$ with respect to w yields

$$\Phi^e(w) = -\frac{e}{2}\left[\hat{\mathbf{E}}_1 \cosh w + \frac{1}{2}\hat{\mathbf{E}}_2 \cosh(2w) + \right.$$

$$\left. \sum_{m=2}^{M-1} \hat{\mathbf{E}}_{m+1}\left(\frac{1}{m+1} \cosh[(m+1)w] - \frac{1}{m-1} \cosh[(m-1)w] \right) \right].$$
(4.65)

The potential Φ^{en} and Φ^{es} can be obtained from this complex function by the following formulae and with $\hat{\mathbf{E}}_m = \hat{E}_m^n$ or $\hat{\mathbf{E}}_m = i\hat{E}_m^s$, respectively. Thus

$$\Phi^{en} = \mathrm{Im}\left[\Phi^e\right] \quad \text{and} \quad \Phi^{es} = \mathrm{Im}\left[i\Phi^e\right].$$
(4.66)

The functions for the field of the normal components can be obtained from $\hat{\mathbf{B}}^e(\mathbf{w})$

$$B_\psi^{en} = \frac{1}{h_t}\mathrm{Re}\left[\hat{\mathbf{B}}^e\right] \quad \text{and} \quad B_\eta^{en} = \frac{1}{h_t}\mathrm{Im}\left[\hat{\mathbf{B}}^e\right],$$
(4.67)

setting $\hat{\mathbf{E}}_m = \hat{E}_m^n$. Similarly the functions for the skew component are then obtained by

$$B_\psi^{es} = \frac{1}{h_t}\mathrm{Re}\left[\hat{\mathbf{B}}^e\right] \quad \text{and} \quad B_\eta^{es} = \frac{1}{h_t}\mathrm{Im}\left[\hat{\mathbf{B}}^e\right],$$
(4.68)

with $\hat{\mathbf{E}}_m = i\hat{E}_m^s$.

As $\cosh[(m-1)w]$ for $m = 1, 2, \ldots$ is an orthogonal system, it can be compared to the different expressions of (4.24) and (4.64) for each multipole. Therefore, a relation between B_y, B_x and B_η, B_ψ can be deduced for each multipole $\hat{\mathbf{E}}_m$ by

$$\left(B_\psi^e + i B_\eta^e\right) = \frac{e \sinh[\eta + i\psi]}{h_t}\left(B_y + i B_x\right).$$
(4.69)

This expression reflects once again that the fields are related to each other by the tangents of the elliptic coordinates (3.24), as one can rewrite these tangents by

Fig. 4.5 Direction of the field components for $\eta = 0$

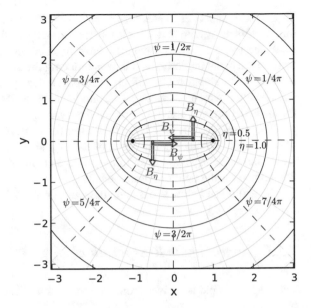

$$(q_\psi + \mathrm{i}\,q_\eta) = \frac{e\,\sinh[\mathbf{w}]}{h_t}\,(q_y + \mathrm{i}\,q_x) = \frac{\sinh[\mathbf{w}]}{\sqrt{\cosh^2 \eta - \cos^2 \psi}}\,(q_y + \mathrm{i}\,q_x), \quad (4.70)$$

with q_x and q_y the tangents for the Cartesian coordinates. For $\eta = 0$ the fraction evaluates to $\pm\mathrm{i}$. This behaviour can be understood studying only the first term $\mathbf{E_1}$. $\mathbf{E_1}$ generates a pure normal or pure skew dipole (Fig. 4.5). Thus the field is independent of x or y. The same multipole generates B_η and B_ψ and must give the field with the same properties. The tangent $\vec{q}_\psi$ or B_ψ points to $-x$ on the upper half plane ($y > 0$) and to x on the lower half plane. So for a pure normal dipole $\hat{E}_1 = 1$ only a field B_η is generated for $\eta = 0$, because (4.70) evaluates to i for $\psi = 0 \ldots \pi$. The tangent q_η points downwards on the lower half and upwards on the upper half. Similar the fraction $e\,\sinh[\mathbf{w}]/h_t$ in (4.70) yields $\pm\mathrm{i}$ for the upper and the lower half. So the tangents times the fraction still give a continuous field. Similar reasoning applies for B_ψ.

4.3 Toroidal Circular Multipoles

For a given beam rigidity an increased magnetic field allows reducing the overall machine size if no dissipative effects (e.g. synchrotron radiation) hinder this approach. The design and development of a new machine is supported by extensive computational studies given the ever increasing capacities of computers.

A particle beam, deflected in a strong dipole, will thus have a significant offset to the axis of a cylinder, which is one of the axes of cylindrical circular coordinates.

Toroidal circular coordinates (see Sect. 3.3.2) are just natural for this problem, as this coordinate system matches closely the curved particle beam path within a dipole field (see Fig. 3.3). Curved magnets have been commonly used, as their curvature follows the beam trajectory and allows reducing the aperture. These coordinates are the obvious candidate for describing the dipole fields of machines with round aperture and a sagitta, which has to be taken into account. SIS300, a machine design made for the FAIR project, e.g. has a projected curvature radius of ≈ 52.6 and it's dipole is of a length of 7 m [24, 25]. This gives a sagitta of ≈ 116 mm, but an beam aperture of only 90 m is foreseen. The value of the sagitta is larger than the beam aperture, which shows clearly that a description based on cylindrical circular multipoles is not first choice for the field of these magnets.

The text given below is based on the work published in [7–9, 15, 20–22].

4.3.1 Approximate R-Separation

The magnetic field is derived from the potential Φ^t. The potential equation is given by

$$\Delta \Phi^t = \left[\frac{\partial^2}{\partial \rho^2} + \frac{1}{\rho} \frac{\partial}{\partial \rho} + \frac{1}{\rho^2} \frac{\partial^2}{\partial \vartheta^2} + \frac{\varepsilon}{h} \left(\cos \vartheta \, \frac{\partial \Phi}{\partial \rho} - \frac{\sin \vartheta}{\rho} \frac{\partial \Phi}{\partial \vartheta} \right) \right] \Phi^t = 0$$

(4.71)

using the Laplacian (3.39) for the local toroidal coordinates. Substituting Φ^t by $\sqrt{h} \, \Phi^t$, with h the metric element of the local toroidal coordinates (3.33), and inserting it into the equation above yields

$$h^{-1/2} \left[\frac{\partial^2}{\partial \rho^2} + \frac{1}{\rho} \frac{\partial}{\partial \rho} + \frac{1}{\rho^2} \frac{\partial^2}{\partial \vartheta^2} + \frac{\varepsilon^2}{h^2} \right] \left(h^{1/2} \, \Phi^t \right) = 0.$$

(4.72)

If ε is small and thus the term ε^2 / h^2 can be neglected (see Sect. 4.3.4) this equation is the same as the one for a straight circular cylinder. This differential equation is described in textbooks (e.g. [23]) and the solutions given there can be used. The problem is considered to be independent of the angle φ. Therefore the same solutions as for 2D polar coordinates are expected. The main effort has now to be devoted to calculate the basis functions. The ansatz is given by

$$\Phi^t_m = h^{-1/2} \, \rho^{|m|} \, e^{im\vartheta} + \mathcal{O}\left(\varepsilon^2 \right) .$$

(4.73)

The inverse coefficient h is developed as a series

$$\frac{1}{\sqrt{h}} = 1 - \frac{1}{2} \rho \cos \left(\vartheta \right) \varepsilon + \frac{3}{8} \rho^2 \cos^2 \left(\vartheta \right) \varepsilon^2 + \mathcal{O}\left(\varepsilon^3 \right)$$

(4.74)

and truncated for all $\mathcal{O}\left(\varepsilon^2\right)$. The term $\rho \cos \vartheta$ equals x / R_{Ref}. Hence one can rewrite the previous equation as

$$S := \frac{1}{\sqrt{h}} = 1 - \frac{1}{2} \varepsilon \frac{x}{R_{\text{Ref}}} + \mathcal{O}\left(\varepsilon^2\right). \tag{4.75}$$

4.3.2 The Potential

Using the complex representation of $\cos \vartheta$ the ansatz yields

$$\Phi_m^t = \frac{1}{m} \frac{1}{\sqrt{h}} R_{|m|}(\rho) e^{im\theta} = \frac{1}{m} \left[\rho^{|m|} e^{im\vartheta} - \frac{\varepsilon}{4} \rho^{|m|+1} \left(e^{i(m+1)\vartheta} + e^{i(m-1)\vartheta} \right) \right],$$
$$\tag{4.76}$$

adding the term $1/m$ to be consistent with the cylindrical circular multipoles. The potential Φ_m^t shall be transferred to the local Cartesian coordinates (i.e. the system local at any given point φ, the toroidal angle)

$$\frac{x}{R_{\text{Ref}}} = \rho \cos \vartheta, \quad \frac{y}{R_{\text{Ref}}} = \rho \sin \vartheta; \quad \mathbf{z} = x + iy. \tag{4.77}$$

The complex potential is given by

$$\Phi_m^t = \frac{1}{m} \left\{ \underbrace{\left(\frac{\mathbf{z}}{R_{\text{Ref}}} \right)^{|m|}}_{T_0} - \frac{\varepsilon}{4} \left[\underbrace{\left(\frac{\mathbf{z}}{R_{\text{Ref}}} \right)^{|m|+1}}_{T_1} + \underbrace{\frac{|\mathbf{z}|^2}{R_{\text{Ref}}^2} \left(\frac{\mathbf{z}}{R_{\text{Ref}}} \right)^{|m|-1}}_{T_2} \right] \right\}. \tag{4.78}$$

The different terms are denoted by T_0, T_1 and T_2, which will be required later on. The term T_0 is the ansatz for the circular cylindrical multipoles (see also (3.16)), while T_1 is shifted up by $m + 1$, thus $T_{1m} = T_{0m+1}$. So for $m = 1$ T_1 is the potential of a pure quadrupole while T_0 is the potential of a pure dipole. The component $|\mathbf{z}|^2$ shows that the above equation is not an complex analytic function.

A scalar product can be defined by

$$\left(\Phi_m^t, \Phi_k^t \right) := \int_{-\pi}^{\pi} \Phi_m^{t\,*}(\rho, \vartheta) \, \Phi_k^t(\rho, \vartheta) \, h \, d\vartheta =$$

$$= \frac{1}{m^2} \int_{-\pi}^{\pi} \rho^m \rho^k \, e^{i(k-m)\vartheta} \, d\vartheta =$$

$$= \frac{1}{m^2} \rho^{2m} \, 2\pi \, \delta_{mk}. \tag{4.79}$$

Hence the basis functions of Φ_m^t are orthogonal. The potential can thus be described by

$$\Phi^t(\rho, \vartheta) = \sum_{m=-\infty}^{\infty} \tau_m \, \Phi_m^t(\rho, \vartheta) \qquad (4.80)$$

using

$$\tau_m = \frac{m^2}{2\pi} \left(\Phi_m^t(1, \vartheta), \Phi^t(1, \vartheta) \right) \qquad (4.81)$$

to compute the expansion coefficients on the reference radius $\rho = 1$. Toroidal multi-poles accurate to second order in ε are given in [9], but not repeated here as neither further results nor conclusions have been derived from these. Readers requiring a concise treatment of the magnetic field or potential of magnets, whose geometry gives a larger ε, are referred to this paper as a starting point.

4.3.3 Real Basis Vector Field in Local Toroidal Coordinates

In Sect. 4.3.2 a potential was derived using solutions in complex notation. Similarly to the basis functions of the potential basis field vectors can be constructed using

$$\Phi_m^{t\alpha}(x, y) = S \, \Phi_m^{t\alpha}(x, y) = \Phi_m^{C\alpha}(x, y) - \frac{1}{2} \varepsilon \frac{x}{R_{\text{Ref}}} \Phi_m^{C\alpha}(x, y); \quad \alpha = n, s \quad (4.82)$$

inserting (4.11) or (4.13) for $\Phi^{C\alpha}$. The constant part of the sum, adding no contribution in the cylindrical circular case, gives a contribution here

$$\Phi_0^{t\alpha}(x, y) = S \, C_0 = C_0 - \frac{1}{2} \varepsilon \frac{x}{R_{\text{Ref}}} C_0. \qquad (4.83)$$

But it is omitted for all calculations below, as the toroidal approximation shall give the same result for

$$\lim_{\varepsilon \to 0} \Phi_0^{t\alpha}. \qquad (4.84)$$

If the term $\Phi_0^{t\alpha}(x, y)$ were included, it would add a skew dipole $\propto \varepsilon$, so if experience will show that this term is required for accelerator fields, it could be included. The expansion coefficients are introduced in the field expansion

$$\vec{B}^t(x, y) = \sum_{m=1}^{M} \left[R_m \, \vec{T}_m^n(x, y) + S_m \, \vec{T}_m^s(x, y) \right]. \qquad (4.85)$$

It is convenient to define a normalised potential function $\hat{\Phi}_m^C(x)$ and a normalised field function $B_m^C(x)$ for each multipole with label m. The normalised potential starts from a definition of the general m^{th} multipole corresponding to the definition (4.2) with

$$\Phi_m^C(\mathbf{z}) = -\frac{1}{m} [B_m + iA_m] \left(\frac{\mathbf{z}}{R_{\text{Ref}}}\right)^m. \tag{4.86}$$

The operator Q

$$Q_{\text{ns}} = (B_m + iA_m \rightarrow 1) \tag{4.87}$$

transformers the general multipols potential into the normalised potential

$$\hat{\Phi}_m^C(\mathbf{z}) = Q_{\text{ns}} \left(\Phi_m^C(\mathbf{z})\right) = -\frac{1}{m} \left(\frac{\mathbf{z}}{R_{\text{Ref}}}\right)^m. \tag{4.88}$$

The normalised basic field function based on a single multipole term of (4.2) yields thus

$$\mathbf{B_m}(\mathbf{z}) = [B_m + iA_m] \left(\frac{\mathbf{z}}{R_{\text{Ref}}}\right)^m = -R_{\text{Ref}} \frac{d\Phi_m^C(\mathbf{z})}{d\mathbf{z}} \tag{4.89}$$

which gives the basic multipole field

$$\hat{\mathbf{B}}_m^C(\mathbf{z}) = -R_{\text{Ref}} \frac{d}{d\mathbf{z}} \hat{\Phi}_m^C(\mathbf{z}) = \left(\frac{\mathbf{z}}{R_{\text{Ref}}}\right)^{m-1}. \tag{4.90}$$

The operator Q_n transforms general expressions to the corresponding normal expressions

$$Q_{\text{n}} = (B_m \rightarrow 1, A_m \rightarrow 0). \tag{4.91}$$

The operator Q_s transforms general expressions to the corresponding skew expressions

$$Q_{\text{s}} = (B_m \rightarrow 0, A_m \rightarrow 1). \tag{4.92}$$

The real and skew toroidal basis functions are then derived by applying the gradient to

$$\vec{T}_m^{C\alpha}(x, y) = -R_{\text{Ref}} \nabla \hat{\Phi}_m^{t\alpha}(x, y), \tag{4.93}$$

which can be further expanded by the differentiation rules to

$$\vec{T}_m^\alpha(x, y) = -R_{\text{Ref}} \nabla \hat{\Phi}_m^{C\alpha}(x, y) + R_{\text{Ref}} \frac{1}{2} \varepsilon \nabla \left(\frac{x}{R_{\text{Ref}}} \hat{\Phi}_m^{C\alpha}(x, y)\right), \quad \alpha = n, s; \tag{4.94}$$

$$= R_{\text{Ref}} \begin{pmatrix} -\frac{\partial \hat{\Phi}_m^{C\alpha}(x,y)}{\partial x} + \frac{1}{2}\varepsilon \frac{x}{R_{\text{Ref}}} \frac{\partial \hat{\Phi}_m^{C\alpha}(x,y)}{\partial x} + \frac{1}{2}\varepsilon \frac{1}{R_{\text{Ref}}} \hat{\Phi}_m^{C\alpha}(x, y) \\ -\frac{\partial \hat{\Phi}_m^{C\alpha}(x,y)}{\partial y} + \frac{1}{2}\varepsilon \frac{x}{R_{\text{Ref}}} \frac{\partial \hat{\Phi}_m^{C\alpha}(x,y)}{\partial y} \end{pmatrix} \tag{4.95}$$

$$= \hat{\vec{B}}(x, y) \left(1 - \frac{1}{2} \varepsilon \frac{x}{R_{\text{Ref}}}\right) + \frac{1}{2} \varepsilon \frac{1}{R_{\text{Ref}}} \hat{\Phi}_m^{C\alpha}(x, y)\vec{i}_x. \tag{4.96}$$

The real normalised potential functions may be obtained from the complex basis functions

$$\hat{\Phi}_m^{Cn}(x, y) = \text{Im}\left(Q_n(\Phi_m^C(z))\right) = \text{Im}\left(\hat{\Phi}_m^C(z)\right), \tag{4.97}$$

$$\hat{\Phi}_m^{Cs}(x, y) = \text{Im}\left(Q_s(\Phi_m^C(z))\right) = \text{Im}\left(i\hat{\Phi}_m^C(z)\right), \tag{4.98}$$

with

$$Q_n\left[\Phi_m^C(z)\right] = \hat{\Phi}_m^C(z), \quad \text{and} \quad Q_s\left[\Phi_m^C(z)\right] = i\hat{\Phi}_m^C(z). \tag{4.99}$$

The underived scalar potential occurs in the last terms of (4.95) and (4.96). Thus it is part of the vectorial basis function. This prohibits a straightforward complex analytic description. Nevertheless this potential can be replaced by a field term. Comparing (4.88) to (4.90) gives

$$\hat{\Phi}_m^C(z) = -\frac{1}{m}\frac{z}{R_{\text{Ref}}}\left(\frac{z}{R_{\text{Ref}}}\right)^{m-1} = -\frac{1}{m}\frac{z}{R_{\text{Ref}}}\hat{B}_m^C(z) \tag{4.100}$$

and thus

$$\hat{\Phi}_m^{Cn}(x, y) = \text{Im}\left(Q_n\left[\Phi_m^C(z)\right]\right) = -\text{Im}\left(\frac{1}{m}\frac{z}{R_{\text{Ref}}}B_m^C(z)\right), \tag{4.101}$$

$$\hat{\Phi}_m^{Cs}(x, y) = \text{Im}\left(Q_s\left[\Phi_m^C(z)\right]\right) = -\text{Im}\left(i\frac{1}{m}\frac{z}{R_{\text{Ref}}}B_m^C(z)\right). \tag{4.102}$$

The total magnetic induction may be represented by (4.85)

$$\vec{B}^t(x, y) = \sum_{m=1}^M \left[R_m\,\vec{T}_m^n(x, y) + S_m\,\vec{T}_m^s(x, y)\right].$$

Now the scalar product is defined by

$$\left(\vec{A} \cdot \vec{B}\right) := \int_{-\pi}^{\pi} \left(\left(A_x B_x + A_y B_y\right)h\right)_{\rho=1} d\vartheta \tag{4.103}$$

$$= \int_{-\pi}^{\pi} \left(\left(A_\rho B_\rho + A_\vartheta B_\vartheta\right)h\right)_{\rho=1} d\vartheta. \tag{4.104}$$

The different basic vector fields are not orthogonal. For a given field $\vec{B}(x, y)$ the expansion coefficients R_m, S_m must be found by solving the following system

$$\left(\vec{T}_k^n \cdot \vec{B}\left(\rho\cos\vartheta, \rho\sin\vartheta\right)\right) = \sum_{m=1}^{M} \left(R_m\left(\vec{T}_k^n \cdot \vec{T}_m^n\right) + S_m\left(\vec{T}_k^n \cdot \vec{T}_m^s\right)\right),$$

$$\left(\vec{T}_k^s \cdot \vec{B}\left(\rho\cos\vartheta, \rho\sin\vartheta\right)\right) = \sum_{m=1}^{M} \left(R_m\left(\vec{T}_k^s \cdot \vec{T}_m^n\right) + S_m\left(\vec{T}_k^s \cdot \vec{T}_m^s\right)\right).$$

$$k = 1, 2, ..., M; \qquad (4.105)$$

In consequence all their values will change every time whenever the number of coefficients, M + M, is changed. The orthogonality of the $\vec{T}_m^\alpha$ to the $\vec{T}_k^\beta$ can be investigated in Green's first theorem [9]. It turns out that every normal basic function is orthogonal to each skew function. But the normal functions are not orthogonal among each other; nor the skew. The error is of the order of ε.

4.3.3.1 Complex Representation of the Basic Vector Field. 1st Approach

A complex representation of the field allows obtaining a closed expression for the magnetic field introducing basis functions $\mathbf{T_m^n}$, $\mathbf{T_m^s}$ which fulfil

$$\mathbf{B^t}(\mathbf{z}) = \sum_{m=1}^{M} \left[R_m\, \mathbf{T_m^n}(\mathbf{z}) + i\, S_m\, \mathbf{T_m^s}(\mathbf{z})\right]. \qquad (4.106)$$

A comparison of (4.85)–(4.106) shows that $\mathbf{T_m^n}$ and $\mathbf{T_m^s}$ are related to $\vec{T}_m^n$, $\vec{T}_m^s$ by

$$\vec{T}_m^n(x, y) := \text{Re}\left[\mathbf{T_m^n}(\mathbf{z})\right]\vec{i}_y + \text{Im}\left[\mathbf{T_m^n}(\mathbf{z})\right]\vec{i}_x, \qquad (4.107)$$

$$\vec{T}_m^s(x, y) := \text{Re}\left[i\mathbf{T_m^s}(\mathbf{z})\right]\vec{i}_y + \text{Im}\left[i\mathbf{T_m^s}(\mathbf{z})\right]\vec{i}_x, \qquad (4.108)$$

with $\vec{i}_x$ and $\vec{i}_y$ the unity vectors in x and y. Furthermore the complex functions can be expressed by $\vec{T}_m^n$ and $\vec{T}_m^s$

$$\mathbf{T_m^n}(\mathbf{z}) = \left(\vec{T}_m^n\left(\text{Re}\left(\mathbf{z}\right), \text{Im}\left(\mathbf{z}\right)\right) \cdot \vec{i}_y\right) + \left(\vec{T}_m^n\left(\text{Re}\left(\mathbf{z}\right), \text{Im}\left(\mathbf{z}\right)\right) \cdot \vec{i}_x\right)i \qquad (4.109)$$

$$\mathbf{T_m^s}(\mathbf{z}) = \left(\vec{T}_m^s\left(\text{Re}\left(\mathbf{z}\right), \text{Im}\left(\mathbf{z}\right)\right) \cdot \vec{i}_y\right) + \left(\vec{T}_m^s\left(\text{Re}\left(\mathbf{z}\right), \text{Im}\left(\mathbf{z}\right)\right) \cdot \vec{i}_x\right)i \qquad (4.110)$$

Similarly terms for $\vec{T}_m^n(\rho, \vartheta)$ and $\vec{T}_m^s(\rho, \vartheta)$ can be defined.

The different terms $\mathbf{T_m^n}$ and $\mathbf{T_m^s}$ are derived below. One can see from (4.94) that the basis functions can be derived from the potential of the cylindrical circular multipoles (4.11) and (4.13). Now concise equations of the field expansions shall be derived for each multipole m. In a first step the different terms in (4.95) will be expressed by a single expression, which will in turn then be combined to a common expression.

The expressions of (4.95) consist of terms for B_y and B_x or the potentials of these fields namely Φ_m^{Cn} and Φ_m^{Cs} for the normal and skew components. A few short cuts are now introduced which allow expressing the derivation in short form in the following.

The complex field vectors may be expressed in the following way

$$\hat{\mathbf{B}}_m^n = Q_n \left(B_y + i B_x \right) = B_y^n + i B_x^n = Q_n \left[\mathbf{B}_m^C(\mathbf{z}) \right] = \hat{\mathbf{B}}_m^C(\mathbf{z}) \tag{4.111}$$

$$B_y^n(\mathbf{z}) = \text{Re}\left[\hat{\mathbf{B}}_m^C \right] \qquad B_x^n(\mathbf{z}) = \text{Im}\left[\hat{\mathbf{B}}_m^C \right] \tag{4.112}$$

$$\hat{\mathbf{B}}_m^s = Q_s \left(B_y + i B_x \right) = B_y^s + i B_x^s = Q_s \left[\mathbf{B}_m^C(\mathbf{z}) \right] = i\hat{\mathbf{B}}_m^C(\mathbf{z}) \tag{4.113}$$

$$\hat{B}_y^s(\mathbf{z}) = \text{Re}\left[i\hat{\mathbf{B}}_m^C \right] \qquad \hat{B}_x^s(\mathbf{z}) = \text{Im}\left[i\hat{\mathbf{B}}_m^C \right] \tag{4.114}$$

As $\mathbf{z} = x + iy$ one can use the expressions above to give short forms for the terms in (4.95). Using the expressions above the different terms of (4.95) are given by

$$\vec{T}_m^n(x, y) = \begin{pmatrix} \text{Im}\left[\hat{\mathbf{B}}_m^C \right] - \frac{1}{2}\,\varepsilon\,\frac{x}{R_{\text{Ref}}}\text{Im}\left[\hat{\mathbf{B}}_m^C \right] + \frac{1}{2}\,\varepsilon\text{Im}\left[\hat{\mathbf{\Phi}}_m^C \right] \\ \text{Re}\left[\hat{\mathbf{B}}_m^C \right] - \frac{1}{2}\,\varepsilon\,\frac{x}{R_{\text{Ref}}}\text{Re}\left[\hat{\mathbf{B}}_m^C \right] \end{pmatrix} \tag{4.115}$$

for the normal components. Similarly one obtains for the skew components

$$\vec{T}_m^s(x, y) = \begin{pmatrix} \text{Im}\left[i\hat{\mathbf{B}}_m^C \right] - \frac{1}{2}\,\varepsilon\,\frac{x}{R_{\text{Ref}}}\,\text{Im}\left[i\hat{\mathbf{B}}_m^C \right] + \frac{1}{2}\,\varepsilon\text{Im}\left[i\hat{\mathbf{\Phi}}_m^C \right] \\ \text{Re}\left[i\hat{\mathbf{B}}_m^C \right] - \frac{1}{2}\,\varepsilon\,\frac{x}{R_{\text{Ref}}}\,\text{Re}\left[i\hat{\mathbf{B}}_m^C \right] \end{pmatrix}. \tag{4.116}$$

Equations (4.115) and (4.116) shall now be combined such $\mathbf{T}^n$ and $\mathbf{T}^s$ can be expressed as terms of $\mathbf{z}$. The first terms in "the first column" of (4.115) and (4.116) are straightforward to combine, as these are the definitions of the cylindric circular multipoles and hence will give

$$\left(\frac{\mathbf{z}}{R_{\text{Ref}}} \right)^{n-1}. \tag{4.117}$$

as first term for $\mathbf{T}^n$ and $\mathbf{T}^s$. x is expressed by

$$x = \text{Re}\,[\mathbf{z}], \tag{4.118}$$

which is a real variable, thus the second column can be combined to

$$\frac{1}{2}\varepsilon\,\frac{\text{Re}\,[\mathbf{z}]}{R_{\text{Ref}}} \left(\frac{\mathbf{z}}{R_{\text{Ref}}} \right)^{m-1}. \tag{4.119}$$

The last term reflects the potential. Each term of the complex potential (4.2) relates to the terms of the complex field (4.8) by

$$\hat{\Phi}_m^C = -\frac{1}{m}\frac{\mathbf{z}}{R_{\text{Ref}}}\hat{\mathbf{B}}_m^C, \tag{4.120}$$

So the last term can be expressed by the relation

$$\text{Im}\left[\hat{\Phi}^C\right] = -\frac{1}{m}\text{Im}\left[\frac{\mathbf{z}}{R_{\text{Ref}}}\hat{\mathbf{B}}^C\right] \quad \text{and} \quad \text{Im}\left[i\hat{\Phi}^C\right] = -\frac{1}{m}\text{Im}\left[i\frac{\mathbf{z}}{R_{\text{Ref}}}\hat{\mathbf{B}}^C\right]. \tag{4.121}$$

This term contributes only to the field B_x. The basis function $\mathbf{T}^n$ is multiplied with R_m, thus this term has to be multiplied with i for $\mathbf{T}^n$. For $\mathbf{T}^s$ it will be multiplied with iS_m so it is not required to multiply it with i. Using $\text{Im}\left[i\Phi^{Cn}\right] = \text{Re}\left[\Phi^{Cn}\right]$, $\mathbf{T}^n$ and $\mathbf{T}^s$ are given by

$$\begin{pmatrix} \mathbf{T}_m^n(\mathbf{z}) \\ \mathbf{T}_m^s(\mathbf{z}) \end{pmatrix} = \left(\frac{\mathbf{z}}{R_{\text{Ref}}}\right)^{m-1}\left(1 - \varepsilon\frac{1}{2}\frac{\text{Re}(\mathbf{z})}{R_{\text{Ref}}}\right) - \varepsilon\frac{1}{2m}\begin{pmatrix} i\text{Im}\left[\left(\frac{\mathbf{z}}{R_{\text{Ref}}}\right)^m\right] \\ \text{Re}\left[\left(\frac{\mathbf{z}}{R_{\text{Ref}}}\right)^m\right] \end{pmatrix}. \tag{4.122}$$

The terms $\mathbf{T}_m^n$ and $\mathbf{T}_m^s$ share a common part, while in the last bracket the upper term is only used for $\mathbf{T}_m^n$ and the lower one only for $\mathbf{T}_m^s$. Substituting $\mathbf{z}/R_{\text{Ref}}$ with $\rho e^{i\vartheta}$ one gets

$$\begin{pmatrix} \mathbf{T}_m^n(\mathbf{z}) \\ \mathbf{T}_m^s(\mathbf{z}) \end{pmatrix} = \rho^{m-1}e^{i(m-1)\vartheta}\left(1 - \varepsilon\frac{1}{2}\rho\cos(\vartheta)\right) - \varepsilon\frac{\rho^m}{2m}\begin{pmatrix} i\sin(m\vartheta) \\ \cos(m\vartheta) \end{pmatrix} \tag{4.123}$$

$$= \rho^{m-1}\left(e^{i(m-1)\vartheta} - \frac{\varepsilon\rho}{2}\left[e^{i(m-1)\vartheta}\cos(\vartheta) + \frac{1}{m}\begin{pmatrix} i\sin(m\vartheta) \\ \cos(m\vartheta) \end{pmatrix}\right]\right)$$

in polar representation. Cylindrical circular multipoles are only satisfying the relevant Maxwell equations if $m > 0$ (see e.g. [18]). The same condition is now imposed on m here as otherwise the aforementioned condition will not be fulfilled for $\varepsilon \to 0$. Therefore it is assumed that $m > 0$ holds also for the toroidal circular multipoles.

4.3.3.2 Complex Representation of the Basic Vector Field. 2nd Approach and Interpretation

The basis functions can also be derived by a second approach starting with $\text{Im}\left(\Phi^t\right)$ (see (4.78)) as complex analytic expressions can yield solutions which are more straightforward to interpret. Given that $|\mathbf{z}|$ is part of the ansatz, the calculation can not be performed using complex coordinates safely. Instead the gradient of $\text{Im}\left(\Phi^t\right)$ is performed by two steps:

• At first all coordinates are substituted by their real values. Then the calculations are performed as defined in (4.94).
• Based on the results complex representations are deduced and Eqs. (4.109) and (4.110) are applied. These equations are chosen such that the term, not part of

the perturbation solution (dependent on ε), will produce the basis functions of conventional multipoles.

These terms $\vec{T}_m$ are calculated to high order using a computer algebra system [24]. Based on these results a formula was obtained and checked. Term 2 of (4.78) yields the following equation

$$\begin{pmatrix} T^n_{2m} \\ T^s_{2m} \end{pmatrix} = \left(\frac{z}{R_{Ref}}\right)^2 (1-m)\left(\frac{z}{R_{Ref}}\right)^{m-2} + \frac{2}{R_{Ref}^{m-1}}\begin{pmatrix} y\,\mathrm{Re}\!\left(z^{m-1}i\right) - ix\,\mathrm{Im}\!\left(z^{m-1}\right) \\ -y\,\mathrm{Im}\!\left(z^{m-1}i\right) - ix\,\mathrm{Re}\!\left(z^{m-1}\right) \end{pmatrix},$$
(4.124)

which can be also expressed by

$$\begin{pmatrix} T^n_{2m} \\ T^s_{2m} \end{pmatrix} = \left(\frac{|z|}{R_{Ref}}\right)^2 \frac{(m-1)}{m}\left(\frac{z}{R_{Ref}}\right)^{m-2} + \frac{2\bar{z}}{m\,R_{Ref}}\begin{pmatrix} i\,\mathrm{Im}\!\left((z/R_{Ref})^{m-1}\right) \\ \mathrm{Re}\!\left((z/R_{Ref})^{m-1}\right) \end{pmatrix},$$
(4.125)

with $\bar{z} = x - iy$. For term T_0 one obtains

$$T^n_{0m} = T^s_{0m} = \left(\frac{z}{R_{Ref}}\right)^{m-1}.$$
(4.126)

and for term T_1

$$T^n_{1m} = T^s_{1m} = \frac{m+1}{m}\left(\frac{z}{R_{Ref}}\right)^m.$$
(4.127)

Thus term T_1 is a "feed up", similar to the "feed-down" effect due to translation of the coordinate system (see Sect. 4.1.2). The results can be combined to

$$\begin{pmatrix} T^n_m \\ T^s_m \end{pmatrix} = \left(\frac{z}{R_{Ref}}\right)^{m-1} -$$
$$-\frac{\varepsilon}{4m}\left\{ \left(\frac{z}{R_{Ref}}\right)^{m-2}\left[(m+1)\left(\frac{z}{R_{Ref}}\right)^2 + (m-1)\left|\left(\frac{z}{R_{Ref}}\right)\right|^2\right] + \right.$$
$$\left. +\frac{2\bar{z}}{R_{Ref}}\begin{pmatrix} i\,\mathrm{Im}\!\left((z/R_{Ref})^{m-1}\right) \\ \mathrm{Re}\!\left((z/R_{Ref})^{m-1}\right) \end{pmatrix}\right\},$$
(4.128)

which is a formulation equivalent to (4.122). Substituting (z/R_{Ref}) with $\rho\exp i\vartheta$ it can be further transformed to

$$\begin{pmatrix} T^n_m \\ T^s_m \end{pmatrix} = \rho^{m-1}\left(e^{i(m-1)\vartheta} - \right.$$
$$\left. -\frac{\rho\varepsilon}{4m}\left\{ e^{i(m-2)\vartheta}\left[(m+1)e^{2i\vartheta} + (m-1)\right] + 2e^{-i\vartheta}\begin{pmatrix} e^{i\pi/2}\sin\left((m-1)\,\vartheta\right) \\ \cos\left((m-1)\,\vartheta\right) \end{pmatrix}\right\}\right),$$
(4.129)

and

$$\begin{pmatrix} \mathbf{T}_m^n \\ \mathbf{T}_m^s \end{pmatrix} = \rho^{m-1} e^{i(m-1)\vartheta} \Bigg(1 - $$

$$- \frac{\rho\varepsilon}{4m} \left\{ \left[(m+1)\,e^{i\vartheta} + (m-1)\,e^{-i\vartheta} \right] + 2e^{-im\vartheta} \begin{pmatrix} e^{i\pi/2} \sin\left((m-1)\,\vartheta\right) \\ \cos\left((m-1)\,\vartheta\right) \end{pmatrix} \right\} \Bigg). $$

$$(4.130)$$

The equations above are just reformulations of (4.128) but simplify the interpretation of the functions. The term dependent on ε is considered as a perturbation term or distortion, as this part does not appear in the field expressions for cylindrical circular multipoles. One can see that

- the whole perturbation depends linearly on the offset from the centre circle.
- The perturbation decreases for higher orders of m.
- The first term $\mathbf{T}_{1m}^n$ or $\mathbf{T}_{1m}^s$ corresponds to a feed up, as if a multipole with order $\varepsilon(m+1)/4$ was added. The term $\mathbf{T}_{2m}^n$ or $\mathbf{T}_{2m}^s$ corresponds to a field increasing with the distance from the centre circle.
- The last term, which is different for $\mathbf{T}^n$ and $\mathbf{T}^s$, is rotating against the phase of the field. Finally $\mathbf{T}^n$ is scaled by the imaginary component while $\mathbf{T}^s$ is scaled by the real component of $z/|z|$ in the direction x.

Table 4.4 Potential and basis functions for term T_2

m	Φ	B_x	B_y
Normal			
1	0	0	0
2	$-\dfrac{x^2 y}{R_{Ref}^3} - \dfrac{y^3}{R_{Ref}^3}$	$\dfrac{xy}{R_{Ref}^2}$	$\dfrac{x^2+3y^2}{2R_{Ref}^2}$
3	$-2\dfrac{x^3 y}{R_{Ref}^4} - 2\dfrac{xy^3}{R_{Ref}^4}$	$\dfrac{2}{3}\dfrac{y(3x^2+y^2)}{R_{Ref}^3}$	$\dfrac{2}{3}\dfrac{x(x^2+3y^2)}{R_{Ref}^3}$
4	$-3\dfrac{x^4 y}{R_{Ref}^5} - 2\dfrac{x^2 y^3}{R_{Ref}^5} + \dfrac{y^5}{R_{Ref}^5}$	$\dfrac{xy(3x^2+y^2)}{R_{Ref}^4}$	$\dfrac{3x^4+6x^2y^2-5y^4}{4R_{Ref}^4}$
5	$-4\dfrac{x^5 y}{R_{Ref}^6} + 4\dfrac{xy^5}{R_{Ref}^6}$	$\dfrac{4}{5}\dfrac{y(5x^4-y^4)}{R_{Ref}^5}$	$\dfrac{4}{5}\dfrac{x(x^4-5y^4)}{R_{Ref}^5}$
Skew			
1	$-\dfrac{x^2}{R_{Ref}^2} - \dfrac{y^2}{R_{Ref}^2}$	$2\dfrac{x}{R_{Ref}}$	$2\dfrac{y}{R_{Ref}}$
2	$-\dfrac{x^3}{R_{Ref}^3} - \dfrac{xy^2}{R_{Ref}^3}$	$\dfrac{3x^2+y^2}{2R_{Ref}^2}$	$\dfrac{xy}{R_{Ref}^2}$
3	$-\dfrac{x^4}{R_{Ref}^4} + \dfrac{y^4}{R_{Ref}^4}$	$\dfrac{4}{3}\dfrac{x^3}{R_{Ref}^3}$	$-\dfrac{4}{3}\dfrac{y^3}{R_{Ref}^3}$
4	$-\dfrac{x^5}{R_{Ref}^5} + 2\dfrac{x^3 y^2}{R_{Ref}^5} + 3\dfrac{xy^4}{R_{Ref}^5}$	$\dfrac{5x^4-6x^2y^2-3y^4}{4R_{Ref}^4}$	$\dfrac{xy(-x^2-3y^2)}{R_{Ref}^4}$
5	$-\dfrac{x^6}{R_{Ref}^6} + 5\dfrac{x^4 y^2}{R_{Ref}^6} + 5\dfrac{x^2 y^4}{R_{Ref}^6} - \dfrac{y^6}{R_{Ref}^6}$	$\dfrac{6x^5-20x^3y^2-10xy^4}{5R_{Ref}^5}$	$\dfrac{2}{5}\dfrac{y(-5x^4-10x^2y^2+3y^4)}{R_{Ref}^5}$

The distortions are non harmonic solutions except for the feed up. The last term shows that the field is decreasing with increasing x. Given that the solution was obtained bending the basis functions and thus "the magnet", this is not surprising. A straight air coil dipole magnet which is bent would show similar behaviour, as the current density decreases on the outside but increases at the inside. The rotations follow similar insight, as the field direction should change too (imagine a straight sextupole coil which is bent to a torus).

Please note that these functions are not analytic complex functions. The complex notation is only used here as it yields a more compact expression. The first basis functions for T_2 are given in Table 4.4. The ones for terms T_0 and T_1 are given in Table 4.1. The basis term T_2 is plotted for the first orders of m (see Fig. 4.6 for $m = 1, 2$ and Fig. 4.7 for $m = 3, 4$). Term T_0 is not plotted as its basis functions are the same as for the cylindrical circular harmonics. Similarly Term T_1 is not plotted as it corresponds to $T_{0m+1} = T_{1m}$. Thus T_1 start with the quadrupole term, then the sextupole follows. The plots of term T_2 show clearly that T_2 is not analytic, as the x and y components do not match the circular cylindrical harmonics. In particular the term $T_{2m=2}$ (see Fig. 4.6f) is a good example. It shows a minimum at $z = 0$ as if a source was present.

The field expansion (4.106) now uses different basis functions for the normal and the skew component. Complex calculation can be used for evaluating the field, but it has to been done separately for the two functions using R_m and iS_m adding up the results afterwards.

Beam dynamic calculations use Taylor expansions of the magnetic field B_y and B_x at $x = 0$, $y = 0$. Using the formulae above one can give the Taylor expansions for term T_2 (the others are already Taylor series)

$$\left. \frac{\partial^{m+1} T_{2\,m}^{\mathrm{n}}}{\partial x^{m+1}} \frac{R_{\mathrm{Ref}}^{m+1}}{(m+1)!} \right|_{x=0, y=0} = \frac{m-1}{m} \vec{i}_y \tag{4.131}$$

$$\left. \frac{\partial^{m+1} T_{2\,m}^{\mathrm{s}}}{\partial x^{m+1}} \frac{R_{\mathrm{Ref}}^{m+1}}{(m+1)!} \right|_{x=0, y=0} = \frac{m+1}{m} \vec{i}_x \tag{4.132}$$

The part $\vec{i}_y$ gives the contribution to B_y while the part with $\vec{i}_x$ gives the contribution to B_x. The index $m + 1$ of the partial derivatives denotes that the mth term contributes only to the $m + 1$ derivative. This corresponds to a "feed-up" (similarly as a "feed-down" is given by the translation of multipoles). As the expansion is performed at the origin, only some scaling coefficients are left over. Comparing these results to the results of a Taylor expansion shows that these basis functions are not the terms of a Taylor series. A Taylor expansion for the term m will approximate the field by

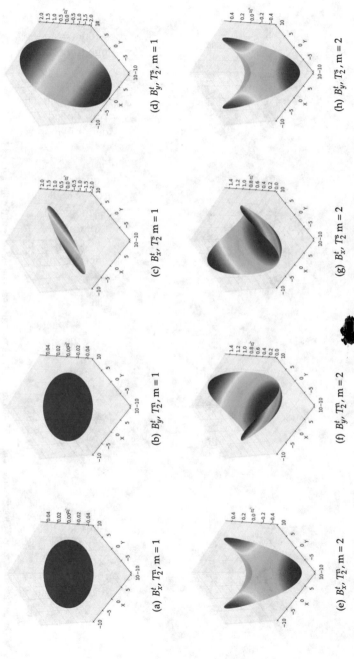

Fig. 4.6 The perturbation term T_2 for the orders $m = 1$ and $m = 2$. The colour code corresponds to the field B_y or B_x

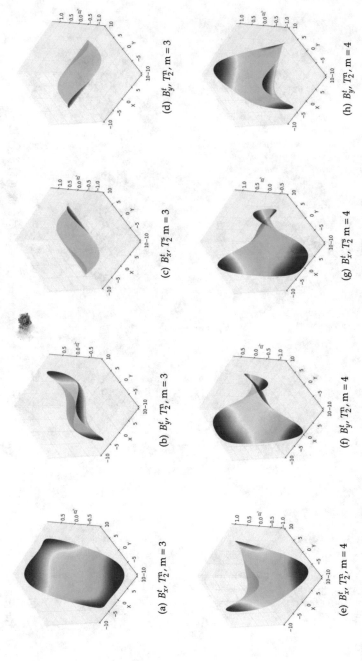

Fig. 4.7 The perturbation term T_2 for the orders $m = 3$ and $m = 4$. The colour code corresponds to the field B_y or B_x

$$B_y \approx R_m \cdot \left.\frac{\partial^{m+1} T^n{}_m}{\partial x^{m+1}} \frac{R_{\text{Ref}}^{m+1}}{(m+1)!}\right|_{x=0,\,y=0} = R_m \left[1 + \frac{\varepsilon}{4}\left(\frac{m+1}{m} + \frac{m-1}{m}\right)\right],$$

(4.133)

$$B_x \approx S_m \cdot \left.\frac{\partial^{m+1} T^s{}_m}{\partial x^{m+1}} \frac{R_{\text{Ref}}^{m+1}}{(m+1)!}\right|_{x=0,\,y=0} = S_m \left[1 + \frac{\varepsilon}{4}\left(\frac{m+1}{m} + \frac{m+1}{m}\right)\right],$$

(4.134)

where the first term is generated by the term T_0, the second by T_1 and the last one by T_2. So any beam dynamic calculations that wants to profit from these new field descriptions has to extend the field description beyond the basis Taylor expansion coefficients. This is once more a consequence of the fact that the term $|\mathbf{z}|$, which is not a complex analytic function, is part of the potential (4.78).

4.3.4   Approximation Error of the Differential Equation

The potential (see (4.2), (4.11), (4.14)) can be given by

$$\Phi_r = \rho^m \left[A_m \cos(m\vartheta) + B_m \sin(m\vartheta)\right],$$

(4.135)

with $\rho = r/R_{\text{Ref}}$. Then the various components of (4.72) are given by

$$\frac{\partial^2 \Phi_r}{\partial \rho^2} = m(m-1)\rho^{m-2} \left[A_m \cos(m\vartheta) + B_m \sin(m\vartheta)\right],$$

(4.136)

$$\frac{1}{\rho}\frac{\partial \Phi_r}{\partial \rho} = m\,\rho^{m-2} \left[B_m \sin(m\vartheta) + A_m \cos(m\vartheta)\right],$$

(4.137)

$$\frac{1}{\rho^2}\frac{\partial^2 \Phi_r}{\partial \vartheta^2} = -m^2\,\rho^{m-2} \left[B_m \sin(m\vartheta) + A_m \cos(m\vartheta)\right].$$

(4.138)

These multipoles are to be used to describe the field within a dipole. Thus B_1 is in the order of $1\,\text{T}$ and ε^2 is compared to this value. ε^2 is $\approx 0.6 \cdot 10^{-6}$ for SIS100. Thus the value can be neglected for SIS100. Similar arguments are applicable for the higher order multipoles as the field description of SIS100 does not need to be better than 10^{-6} or $1\,\text{ppm}$.

4.4   Toroidal Elliptical Multipoles

Finally the deformed circle and "bent" cylinder are combined local toroidal elliptical coordinates. The results given here are based on [20–22, 25].

The local toroidal elliptical coordinates are given in Sect. 3.3.3 together with the appropriate Laplace operator. Thus the potential equation in local toroidal elliptical coordinates is given by

$$\frac{1}{\cosh(2\bar{\eta}) - \cos(2\bar{\psi})} \left[\frac{\partial^2}{\partial\bar{\eta}^2} + \frac{\partial^2}{\partial\bar{\psi}^2} - \frac{\bar{\varepsilon}}{\bar{h}} \left(\sinh\bar{\eta} \cos\bar{\psi} \frac{\partial}{\partial\bar{\eta}} + \cosh\bar{\eta} \sin\bar{\psi} \frac{\partial}{\partial\bar{\eta}} \right) \right] \bar{\Phi} = 0.$$
(4.139)

The constant factor can be omitted. The same approach as for the toroidal circular multipoles is used for the local toroidal multipoles (see Sect. 4.3): "bending the basis functions" substituting Φ by $\sqrt{\bar{h}}\,\Phi$. This yields

$$\frac{1}{\sqrt{\bar{h}}} \left[\frac{\partial^2}{\partial\bar{\eta}^2} + \frac{\partial^2}{\partial\bar{\psi}^2} - \frac{\bar{\varepsilon}^2}{8\bar{h}^2} \left(\cosh(2\bar{\eta}) - \cos(2\bar{\psi}) \right) \right] \left(\sqrt{\bar{h}}\bar{\Phi} \right) = 0.$$
(4.140)

As above any terms of order $\mathcal{O}\left(\bar{\varepsilon}^2\right)$ are neglected. The remaining equation resembles the differential equation as found for the elliptical cylindrical multipoles (see Sect. 4.2). Thus all results given above can be applied here and only the basis functions are to be calculated. The basis functions of the elliptical cylindrical multipoles are, as given above,

$$\begin{aligned} ce_0 &= 1, \\ ce_m(\bar{\eta}, \bar{\psi}) &= \cosh(m\,\bar{\eta}) \cos(m\,\bar{\psi}), \\ se_m(\bar{\eta}, \bar{\psi}) &= \sinh(m\,\bar{\eta})\,\sin(m\,\bar{\psi}). \end{aligned} \quad m = 1, 2, 3, \ldots$$
(4.141)

The basis functions of the toroidal elliptical multipoles are obtained multiplying the basis functions of the cylindrical solution with the square root of the metric element $\bar{h}$, which yields

$$\bar{\Phi}_{cn}(\bar{\eta}, \bar{\psi}) = \frac{1}{\sqrt{\bar{h}}}\, ce_n(\bar{\eta}, \bar{\psi}) + \mathcal{O}\left(\bar{\varepsilon}^2\right)$$
(4.142)

$$= \mathcal{S}(\bar{\eta}, \bar{\psi}) \cosh(n\,\bar{\eta}) \cos(n\,\bar{\psi}) + \mathcal{O}\left(\bar{\varepsilon}^2\right),$$
(4.143)

$$\bar{\Phi}_{sn}(\bar{\eta}, \bar{\psi}) = \frac{1}{\sqrt{\bar{h}}}\, se_n(\bar{\eta}, \bar{\psi}) + \mathcal{O}\left(\bar{\varepsilon}^2\right)$$
(4.144)

$$= \mathcal{S}(\bar{\eta}, \bar{\psi}) \sinh(n\bar{\eta}) \sin(n\bar{\psi}) + \mathcal{O}\left(\bar{\varepsilon}^2\right).$$
(4.145)

The function $\mathcal{S}(\bar{\eta}, \bar{\psi})$ is the first order approximation of $\sqrt{\bar{h}}$:

$$\mathcal{S}(\bar{\eta}, \bar{\psi}) = \left(1 - \frac{1}{2}\bar{\varepsilon} \cosh(\bar{\eta}) \cos(\bar{\psi}) \right) + \mathcal{O}\left(\bar{\varepsilon}^2\right).$$
(4.146)

Further investigations are in progress. The solutions can not be derived directly from the ones given in Sect. 4.2 as these are based on Cartesian fields with elliptical coordinates.

4.5 Summary

Based on the coordinate systems given in Chap. 3 the multipoles were presented for 4 different coordinate systems:

1. cylindrical circular coordinates,
2. cylindrical elliptical coordinates,
3. toroidal circular coordinates and
4. toroidal elliptical coordinates.

All multipoles share that the field is not expected to vary along one coordinate: along the axis for the cylinder or along the toroidal centre circle (large circle) for the torus. The multipoles of the first system are a Taylor series and given in standard text books and well known. Commonly these are given by powers of z^n. The cylindrical elliptical ones are harmonics of $\cosh(nw)$. The elliptical coefficients can be recalculated to the circular ones using matrices. The coefficients of these matrices are obtained by analytic expressions (see Sect. 4.2.2). These matrices show that only normal circular multipoles couple to normal elliptical ones and skew circular couple to skew elliptical ones. The multipoles for the cylindrical coordinate systems can be represented as complex functions.

The calculations show that in theory the cylindrical and elliptical multipoles shall give equivalent results. The power of the new expression is only revealed if measurement or numeric data have to be processed. For domains with an elliptical boundary elliptical coordinates can be calculated and transformed to circular ones (see Sect. 4.2.2. Different methods have been suggested, which could be used to calculate cylindrical circular coordinates directly (e.g. using Cauchy's residue theorem). The advantage of the approach described here is that it not only presents adapted multipoles, but also shows how the coefficients are to be obtained by calculations using (4.28) next to the conversion matrices.

The toroidal multipoles use a similar solution path: apply a transformation on the original potential, which reduces the differential equation to the cylindrical one, if higher order terms of the curvature can be neglected. Then the cylindrical circular solutions can be used for the mapped equation. Further the transformed basis functions are derived.

The toroidal circular basis functions consist of the original circular term and two correction terms, where the first one is a "feed-up" term, while the later one is not analytic any more. This term has an effect on all basis functions, which can not be retransformed to cylindrical circular functions. This also means that these descriptions can only be applied by beam dynamic codes, which use Taylor series expansions for describing the magnetic field, if the internal method of field description is updated. The Taylor expansion coefficients, given in (4.131) and (4.132), show that due to the bend the field B_y "appears" to be a bit weaker than the field B_x. The basis functions are given for the first orders in Table 4.4. There one can see that the basis functions of the original term are not Taylor series terms. The ratio of curvature $\varepsilon = R_{\text{Ref}}/R_{\text{C}}$, together with the required accuracy of field representation, define if these terms have to be taken into account.

For toroidal elliptical multipoles the approximation of the differential equation was given. Again the inverse aspect ratio of the curvature $\bar{\varepsilon} = \bar{e}/R_C$ together with the field representation accuracy indicate if one has to take the perturbation terms into account. It will be shown in Sect. 8.2.4 that these are not required for the current applications. These multipoles are an interesting enhancement and can be helpful for describing the magnetic field of other already planned accelerators.

The field descriptions, applied in beam dynamics within the Frenet-Serret coordinates, use Taylor series along the particle path. Here one can state:

1. the elliptical ones allow describing the field for an elliptical aperture,
2. the toroidal ones follow the circle curvature and thus the multipoles will follow the trajectory of a particle in a dipole field. Therefore these multipoles are to be used for describing the field homogeneity in a dipole if the particle offset from the straight line will give field variations along the particle's trajectory, which can not be neglected.

References

1. A. Wolski, Maxwell's equations for magnets, in *CERN Accelerator School: Specialised Course on Magnets*, ed. by D. Brandt, vol. CERN-2010-004 (CERN, CERN, 2010), pp. 1–38. Published as CERN Yellow Report, http://cdsweb.cern.ch/record/1158462
2. H. Wiedemann, *Particle Accelerator Physics*. (Springer, Berlin, 2007)
3. A. Devred, M. Traveria, Magnetic field and flux of particle accelerator magnets in complex formalism. *not published*, (1994)
4. P. Schnizer, *Measuring system qualification for LHC arc quadrupole magnets*. PhD thesis, (TU Graz, 2002)
5. P. Schnizer, B. Schnizer, P. Akishin, E. Fischer, Magnetic field analysis for superferric accelerator magnets using elliptic multipoles and its advantages. IEEE Trans. Appl. Supercon. **18**(2), 1605–1608 (2008)
6. P. Schnizer, B. Schnizer, P. Akishin, E. Fischer, Theory and application of plane elliptic multipoles for static magnetic fields. Nucl. Instrum. Methods Phys. Res. Sect. A **607**(3), 505–516 (2009)
7. P. Schnizer, B. Schnizer, P. Akishin, E. Fischer, Plane elliptic or toroidal multipole expansions within the gap of straight or curved accelerator magnets, in *13th International IGTE Symposium*. (Institut für Grundlagen und Theorie der Elektrotechnik, Technische Universität Graz, Austria, September 2008)
8. P. Schnizer, B. Schnizer, P. Akishin, E. Fischer, Theoretical field analysis for superferric accelerator magnets using plane elliptic or toroidal multipoles and its advantages, in *The 11th European Particle Accelerator Conference*. (June 2008), pp. 1773–1775
9. P. Schnizer, B. Schnizer, P. Akishin, E. Fischer, Plane elliptic or toroidal multipole expansions for static fields. Applications within the gap of straight and curved accelerator magnets. Int. J. Comput. Math. Electr. Eng. (COMPEL), **28**(4), (2009)
10. P. Schnizer, B. Schnizer, P. Akishin, E. Fischer, Field representation for elliptic apertures. Technical report, (Gesellschaft für Schwerionenforschung mbH, Planckstraße 1, D-64291 Darmstadt, February, 2007)
11. P. Schnizer, B. Schnizer, P. Akishin, E. Fischer, Field representation for elliptic apertures. Technical report, (Gesellschaft für Schwerionenforschung mbH, Planckstraße 1, D-64291 Darmstadt, January, 2008)
12. E.A. Perevedentsev, A.L. Romanov, Eddy current effect on field multipoles arising in dipole magnets with elliptic and rectangular beam pipe, in *The 11th European Particle Accelerator Conference*. (June 2008), pp. 2383–2385

13. F.R. Peña, G. Franchetti, *Elliptic and circular representation of the magnetic field for SIS100* (Technical report, GSI, 2008)
14. Chad E. Mitchell, Alex J. Dragt, Accurate transfer maps for realistic beam-line elements: Straight elements. Phys. Rev. ST Accel. Beams **13**, 064001 (2010)
15. P. Schnizer, B. Schnizer, E. Fischer, Cylindrical circular and elliptical, toroidal circular and elliptical multipoles, fields, potentials and their measurement for accelerator magnet, (October 2014) *arXiv preprint physics.acc-ph*
16. O.D. Kellog, *Foundations of Potential Theory* (Frederick Ungar Publ. Comp, New York, 1929), p. 213
17. S. Großmann, *Funktionalanalysis I.* (Akademische Verlagsgesellschaft, Frankfurt/Main, 1970), p. 82
18. P.M. Morse, H. Feshbach, *Methods of Theoretical Physics*, (McGraw Hill Book Comp., 1953)
19. S.I. Gradshteyn, I.M. Ryzhik, *Table of Integrals* (Academic Press, Series and Products, 1965)
20. P. Schnizer, B. Schnizer, P. Akishin, E. Fischer, Toroidal circular and elliptic multipole expansions within the gap of curved accelerator magnets, in *14th International IGTE Symposium, Graz, Institut für Grundlagen und Theorie der Elektrotechnik* (Technische Universität Graz, Austria, 2010)
21. P. Schnizer, B. Schnizer, E. Fischer, Magnetic field description in curved accelerator magnets using local toroidal multipoles, in *Proceedings of the IPAC'11, San Sebastián, Spain,* (September 2011), pp. 2154–2156
22. P. Schnizer, B. Schnizer, E. Fischer, A. Mierau, Some comments to magnetic field representation for beam dynamic calculations, in *Proceedings of IPAC2012,* (New Orleans, Louisiana, USA, 2012), pp. 262–264
23. E.A. Kraut, *Fundamentals of Mathematical Physics.* (Dover Publications Inc., 1995)
24. SymPy Development Team, *SymPy: Python library for symbolic mathematics* (2014)
25. P. Schnizer, B. Schnizer, P. Akishin, A. Mierau, E. Fischer, SIS100 dipole magnet optimisation and local toroidal multipoles. IEEE Trans. Appl. Supercon **22**(3), 4001505–4001505 (2012)

Chapter 5
Rotating Coils

Magnetic measurement instruments which have been commonly used for measuring the fields of accelerator magnets are based on the following effects:

- magnetic field induction [1]
- Hall effect [2]
- nuclear magnetic resonance [3]
- flux gates.

While systems based on other effects have been built, these are the ones commonly used. Taylor series expansions allow a description of magnetic fields in a simplified beam dynamics approach, thus a measurement method which renders these coefficients is desirable. The harmonic coil method [4] has been proven to be a reliable method to obtain these coefficients and thus has been applied for obtaining the harmonic coefficients of accelerator magnets during the measurement campaigns on various projects (e. g. HERA at DESY[5], LHC at CERN [6–9]). A detailed overview of the method is given in [10] and further elaborations on the impact of different artefacts are given in [7, 11, 12].

The following discussion is an abbreviation of the original text given in [7] and just reflects the parts required to aid the reader's comprehension of the following chapters. Many conclusions drawn later on are independent of the chosen measurement method but are inherent to the geometry and multipoles used for describing the field.

5.1 Derivation of Coil Probe Geometry Factors

A wire loop rotating around an axis forms a rotating coil probe. In Fig. 5.1 a sketch of such a coil probe is given. The wire loop integrates the field over its area. The induced signal is measured versus time by a data acquisition system. In the following

© Springer International Publishing AG 2017
P. Schnizer, *Advanced Multipoles for Accelerator Magnets*, Springer Tracts in Modern Physics 277, DOI 10.1007/978-3-319-65666-3_5

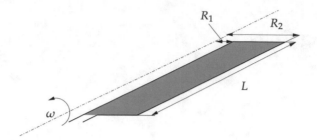

Fig. 5.1 Sketch of a rotating coil. Rotating coils measure the magnetic field based on the induction law. When a coil is rotated a voltage is induced into the wire loop, which is proportional to the field passing the coils surface (indicated by the *grey rectangle*). Beside the field the induced voltage depends on the rotation speed ω the length L, the outer radius R_2 and the inner radius R_1

the required mathematical treatment is given, which allows interpreting the obtained signal and deducing the coefficients of cylindrical circular multipoles.

5.1.1 Complex Potential

During the rotation, rotating coil probes cover the area of a circular cylinder. Thus cylindrical circular coordinates are the standard coordinate system for describing them. Similar to the real potential (see Sect. 4.1) a complex analytic potential $\mathbf{F}(\mathbf{z})$ can be defined by

$$\mathbf{B}(\mathbf{z}) = -\frac{d\mathbf{F}(\mathbf{z})}{d\mathbf{z}} . \tag{5.1}$$

Apart form the sign the complex function $\mathbf{F}$ is a primitive of $\mathbf{B}$. Hence it is regular and analytic. The potential can be expanded in a power series by

$$\mathbf{F}(\mathbf{z}) = -\sum_{n=1}^{\infty} \frac{1}{n} \, \mathbf{C_n} \, \frac{\mathbf{z}^n}{R_{Ref}^{\,n-1}} + \text{const} . \tag{5.2}$$

5.1.2 Magnetic Flux Through a Surface

A coil probe covers a certain area. Now a cylindrical surface Σ parallel to the axis of the magnet z_m and uniform in the axial direction is considered. Γ designates the

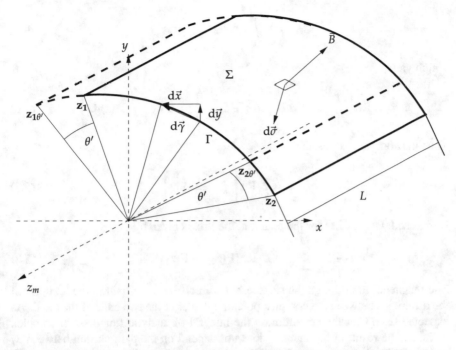

Fig. 5.2 Magnetic flux through a cylindrical surface. A cylindrical surface parallel to the axis z_m indicated by the solid line is turned by an angle θ' to reassemble the dashed surface. L is the length of the surface, z_2, z_1 the original positions of the ends and $z_{2\theta'}$, $z_{1\theta'}$ these positions after the surface has been rotated. Γ represents the intersection arc of the surface Σ and the xy plane. $d\vec{\sigma}$ represents the surface element vector and $\vec{B}$ the direction of the magnetic field

arc at the intersection between Σ and the xy plane. z_1 and z_2 determine the positions of the ends in the complex plane (see Fig. 5.2). The magnetic flux Φ through this surface is described by

$$\Phi = \iint_{\Sigma} B \cdot d\vec{\sigma}, \tag{5.3}$$

with $d\vec{\sigma}$ the surface element vector.

Since the surface is parallel to the axis of the magnet, and since $\vec{B}$ and Σ are uniform along the magnet's axis, one can write

$$\Phi = L \int_{\Gamma} B \cdot \left(\vec{i}_{z_m} \times d\vec{\gamma} \right), \tag{5.4}$$

with L the length of the surface along the z_m-axis, $d\vec{\gamma}$ the arc element vector and $\vec{i}_{z_m}$ the unity vector in direction z_m. The coefficients of the vector $d\vec{\gamma}$ are $(dx, dy, 0)$. The coefficients of $\left(\vec{i}_{z_m} \times d\vec{\gamma} \right)$ are $(-dy, dx, 0)$. Thus the flux is given by

$$\Phi = L \int_{\Gamma} \left(B_y \mathrm{d}x - B_x \mathrm{d}y \right), \qquad (5.5)$$

which yields

$$\Phi = L\,\mathrm{Re}\left[\int_{\Gamma} \left(B_y + \mathrm{i}\,B_x \right) \left(\mathrm{d}x + \mathrm{i}\,\mathrm{d}y \right) \right] = L\,\mathrm{Re}\left[\int_{\Gamma} \mathbf{B(z)}\mathrm{d}\mathbf{z} \right]. \qquad (5.6)$$

Thus the complex expression

$$\Phi = L\,\mathrm{Re}\left[\int_{\mathbf{z_1}}^{\mathbf{z_2}} \mathbf{B(z)}\mathrm{d}\mathbf{z} \right] \qquad (5.7)$$

is obtained. Using (5.1) for the potential the flux is given by

$$\Phi = -L\ \mathrm{Re}\left[\mathbf{F(z_2)} - \mathbf{F(z_1)} \right]. \qquad (5.8)$$

The magnetic flux through the surface Σ is directly proportional to the real part of the difference between the complex potential values at the two ends of the arc Γ. As expected from Cauchy's theorem on the integral of analytic functions in complex variables the result of the integral does not depend on the path chosen between $\mathbf{z_1}$ and $\mathbf{z_2}$. In the above equation $\mathbf{F}$ is replaced by its power series expansion

$$\Phi = L\,\mathrm{Re}\left[\sum_{n=1}^{\infty} \frac{1}{n}\, \mathbf{C_n}\, \frac{\mathbf{z_2}^n - \mathbf{z_1}^n}{R_{Ref}^{\,n-1}} \right]. \qquad (5.9)$$

5.1.3 Magnetic Flux Picked Up by a Rotating Coil

Now it is assumed that the surface Σ represents the surface for all turns of a pick up coil rotating around the axis z_m (i.e. the windings are infinitely thin). The angle θ' describes a rotation of the surface Σ around the axis z_m. $\mathbf{z_2}$ and $\mathbf{z_1}$ are the positions of the extremities of the arc Γ at $\theta' = 0$. So for any angle θ' the location of the ends $\mathbf{z_{1\theta'}}$ and $\mathbf{z_{2\theta'}}$ is described by

$$\mathbf{z_{1\theta'}} = \mathbf{z_1} \exp(\mathrm{i}\theta') \qquad \text{and} \qquad \mathbf{z_{2\theta'}} = \mathbf{z_2} \exp(\mathrm{i}\theta'). \qquad (5.10)$$

Using Eqs. (5.9) and (5.10) the flux Φ seen by a rotating coil is

$$\Phi(\theta') = \mathrm{Re}\left\{ \sum_{n=1}^{\infty} \mathbf{K_n C_n} \exp\left(\mathrm{i}n\theta' \right) \right\}, \qquad (5.11)$$

with $\mathbf{K_n}$ the coil's sensitivity to the nth multipole

$$\mathbf{K_n} = \left(\frac{N_w L R_{\text{Ref}}}{n}\right)\left[\left(\frac{\mathbf{z_2}}{R_{\text{Ref}}}\right)^n - \left(\frac{\mathbf{z_1}}{R_{\text{Ref}}}\right)^n\right]. \tag{5.12}$$

N_w is the number of turns of the coil windings. Equation (5.12) shows that $\mathbf{K_n}$ only depends on the coil geometry.

5.2 Radial Rotating Coil Layout

Equation (5.12) describes the sensitivity of the coil for an arbitrary location of the coil extremities with respect to the rotation axis. Two typical layouts are the *tangential coil* and the *radial coil* (e.g. [7, 10]). The rotating coil probe used here is based on a radial coil probe and is described in the following.

In a radial coil all wires are located in one plane, which passes through the axis of rotation and is parallel to the axis of rotation. In its perfect form it is only sensitive to the tangential component of the magnetic field B_θ (see Fig. 5.3). The sensitivity of such a coil is obtained by replacing $\mathbf{z_2}$ with R_2 and $\mathbf{z_1}$ with R_1 in (5.12)

$$\mathbf{K_n} = \frac{N_w L R_{\text{Ref}}}{n}\left[\left(\frac{R_2}{R_{Ref}}\right)^n - \left(\frac{R_1}{R_{Ref}}\right)^n\right], \tag{5.13}$$

with R_2 the outer diameter and R_1 the inner diameter of the coils windings.

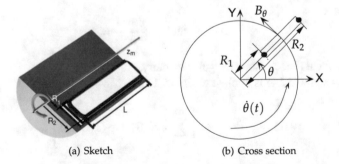

(a) Sketch   (b) Cross section

Fig. 5.3 A radial coil. On the *left* (**a**) a sketch of such a coil and on the *right* (**b**) the cross section is shown. B_θ is the tangential component of the field. R_1 and R_2 are the radii of the windings. In the cross section the axis z_m is perpendicular to the paper plane. The position of the centre of the coil with respect to the x-axis is described by the angle θ and the coil's angular velocity with $\dot{\theta}(t)$

5.3 Voltage Induced in a Rotating Pick Up Coil

The voltage induced by a flux change is given by Faraday's law

$$V = -\frac{d\Phi}{dt} \ . \tag{5.14}$$

A change of flux inside the coil is achieved either by varying the magnetic field (i.e. varying the magnet current) or by rotating the coil inside the magnetic field. Here the second technique, called *rotating coil method*, is described. The angular dependence of the flux on the angular position of the coil is shown in Eq. (5.11). In the following the magnetic field is considered to be independent from time. To calculate the multipoles C_n the flux versus angle is needed. Therefore the measurement is performed in the following way:

- The coil is turned by a motor.
- The voltage induced in the coil is fed to an integrator.
- The integrator is read out by a controller.
- An angular encoder triggers this readout to ensure equidistant readouts. This is needed by the following analysis which is based on a Fourier transform.

In the following this procedure is described mathematically. It is assumed that the N_w turn pick up coil is rotating around the z-axis with some angular velocity. Then the angle θ' at a given time t equals $\theta(t)$ and the angular speed equals its first derivative

$$\theta' = \theta(t) \quad \text{and} \quad \frac{d\theta(t)}{dt} = \dot{\theta}(t). \tag{5.15}$$

In the ideal case

$$\theta' = \omega \cdot t \quad \text{and} \quad \frac{d\theta(t)}{dt} = \omega, \tag{5.16}$$

with ω the ideal (i.e. constant) angular velocity.
Faraday's law (5.14) is applied to Eq. (5.11),

$$V(t) = -\dot{\theta}(t) \ \mathrm{Re}\left\{\sum_{n=1}^{\infty} n\mathbf{K_n C_n} \exp\left(\mathrm{i}n\theta(t)\right)\right\}. \tag{5.17}$$

The voltage is integrated using an integrator

$$\Phi(t) = -\int_{0}^{t} V(t')dt'. \tag{5.18}$$

The angular encoder triggers the readout of the integrator to ensure equally spaced angular steps. Since $\theta(t)$ gives the position of the coil versus time, its inverse function $t = \theta^{-1}(\theta')$ describes the time at which an angle is reached. Thus one can write the flux Φ_i given by the integrator for an angular interval $\theta'_i - \theta'_0$ as

$$\Phi_i = - \int_{\theta^{-1}(\theta'_0)}^{\theta^{-1}(\theta'_i)} V(t)dt. \tag{5.19}$$

θ'_0 is the angle at which the integration starts and θ'_i

$$\theta'_i = \frac{2\pi}{P} i \quad i = 1 \ldots P, \tag{5.20}$$

with P the number of readings per revolution. The flux Φ_i can be further written as

$$\Phi_i = - \int_{t_0}^{t_i} V(t)dt = - \int_{\theta'_0}^{\theta'_i} \frac{1}{\dot\theta} V(\theta)\,d\theta. \tag{5.21}$$

Φ_i corresponds to the value of the integral at t_i. Comparing the last term of (5.21) to (5.17) one finds that Φ_i is independent of the rotating speed of the coil probe thus one obtains

$$\Phi(\theta) = - \ \mathrm{Re}\left\{ \sum_{n=1}^{\infty} \mathbf{K_n C_n} \exp(in\theta) \right\}. \tag{5.22}$$

A discrete Fourier transform is applied to the total readout $\Phi = \{\Phi_i \mid i = 1 \ldots P\}$ of the integrator

$$\Psi = DFT\,[\Phi], \tag{5.23}$$

with Ψ the spectrum of the flux and DFT the discrete Fourier transform. The amplitude of the different harmonics of (5.22) is given by $\mathbf{K_n C_n}$, thus the multipoles are obtained by

$$\mathbf{C_n} = \frac{1}{\mathbf{K_n}} \Psi_n . \tag{5.24}$$

5.4   Compensated Systems

In an accelerator magnet the main harmonic is typically 10^4 times stronger than the other harmonics. This means that the voltage signal induced by the main harmonic is $\approx 10^4$ stronger than the voltage signal induced by the other harmonics. The other harmonics have to be measured with an accuracy to $\approx 10\,\mathrm{ppm}$ or better. A sophisticated

Fig. 5.4 The field coil array
(commonly "harmonic
coil"). One can clearly see
the radial coil layout. All
coils are located on one
plane. The rotation axis
coincides with the central
axis of the cylinder. The coil
E is a spare coil. Drawing
not to scale

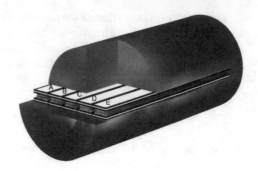

arrangement of coils allows measuring the higher order harmonics with a system
that has a very low sensitivity to the main harmonic. This increases the tolerances
required for the mechanics and electronics [7, 10]. Without this arrangement the
electronics had to measure the induced voltage with a precision of ≈ 2 ppm which
assumes a dynamic range of the measuring channel in the order of 10^6 to cover all
needs (i.e. resolving $10\,\mu V$ on a level of 5 V). This is technically hardly achievable,
therefore the field measurement is split into two parts:

- the *absolute* part. One single coil is used to measure the field strength and its
  angle. The electronic system must guarantee a precision of $\approx 10^{-4}$ to allow the
  field strength measurement with the required precision (1.5×10^{-4}).
- the *compensated* part, which is used to measure the higher order harmonics. Here
  an array of coils is used. They are located on one common body and arranged in
  such a way that a simple summation of the signals allows cancelling or at least
  reducing the dipole and quadrupole component (see Fig. 5.4).

  Due to this reduction the voltage induced by the higher order harmonic has only to
  be measured with a precision of $\approx 10^{-4}$ instead of $\approx 10^{-6}$ (i.e. $10\,\mu V$ on a level of
  50 mV) which allows reducing the requirements to the electronics by two to three
  orders of magnitude.

Only radial coils are considered here (see Fig. 5.4). Coil A is used for the absolute
system as it has the highest sensitivity to the main field. Two different compensation
schemes can be implemented with this coil array. The sensitivity of the coil array
to the dipole harmonic can be reduced connecting the coils $A - C$ and its sensitivity to the quadrupole can be reduced by using $A - B - C + D$ [7]. Coil E is
added to get a symmetric layout. For a perfectly manufactured array the dipole and
quadrupole components would cancel. However small values are left due to small
coil imperfections occurring in the fabrication, due to errors in the geometry of the
coils on the support and due to misalignment of the support with respect to its axis of
rotation. These errors are on the order of micrometers. Therefore the bucking factor
b is defined by

$$b = \left| \frac{\mathbf{K}_{m_{abs}}}{\mathbf{K}_{m_{cmp}}} \right|, \tag{5.25}$$

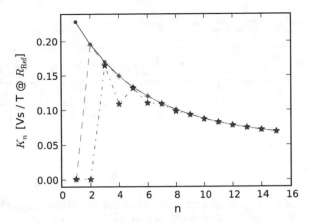

Fig. 5.5 Comparison of the sensitivities $\mathbf{K_n}$ in (Volt seconds per Tesla at the reference Radius) for the absolute (solid *blue line*), the compensated system $A - C$ (*green dashed line*) and the the compensated system $A - B - C + D$ (*red dash dotted line*) system versus the harmonic n for the harmonic probe. The absolute sensitivity shows the typical characteristics of a radial coil. The sensitivity to the dipole or quadrupole component is almost zero for the compensated systems. The sensitivity of the compensated systems nearly matches the sensitivity of coil probe A for the higher order harmonics

with $\mathbf{K_{m_{abs}}}$ the sensitivity of coil A to the main multipole and $\mathbf{K_{m_{cmp}}}$ the sensitivity to the main multipole of the coil array $A - B - C + D$ or $A - C$. This factor has to be measured and is equal to the ratio

$$\frac{V_{m_{abs}}}{V_{m_{cmp}}},\qquad(5.26)$$

with $V_{m_{abs}}$ the voltage induced in the absolute coil by the main multipole and $V_{m_{cmp}}$:

$$V_{m_{cmp}} = V_{m_a} - V_{m_b} - V_{m_c} + V_{m_d},\qquad(5.27)$$

with V_{m_a} the voltage induced in coil A by the main multipole and, for the coils B, C, D respectively. For a dipole system a good bucking factor is in the range of 1000, for a quadrupole system in the range of 100 [13], since the dipole sensitivity is proportional to $\mathbf{z}^1$ and the quadrupole sensitivity is proportional to $\mathbf{z}^2$ (see Eq. (5.12)). The harmonic coil of the system described here has a quadrupole bucking factor of 300. Therefore the electronics of the system described here only needs to resolve $10\,\mu$V on a level of ≈ 15 mV to achieve a precision of 2 ppm. The comparison between the sensitivity of the absolute and compensated system is given in Fig. 5.5.

The mechanical parameters of its five coils probes (named "A" to "E") are given in Table 5.1 [14]. Coil calibration techniques are described in e.g. [15] or [16].

Table 5.1 Mechanical parameters of the coil probes of the array

	A	B	C	D	E
$R_2(mm)$	17.21	10.00	2.78	9.71	16.99
$R_1(mm)$	11.91	4.70	−2.52	4.41	11.69
N	64	64	64	64	64
$L(mm)$	599.3	599.3	599.3	599.3	599.3

References

1. W. Weber, Über die Anwendung der magnetischen Induktion auf Messung der Inclination mit dem Magnetometer. Ann. der Physik. **166**, 209–247 (1853)
2. E.H. Hall, On a new action of the magnet on electric currents. Am. J. Math. **2**, 287–292 (1879)
3. I.I. Rabi, J.R. Zacharias, S. Millman, P. Kusch, A new method of measuring nuclear magnetic moment. Phys. Rev. **53**, 318–318 (Feb, 1938)
4. J.K. Cobb, R.S. Cole, Spectroscopy of quadrupole magnet, in *Proceedings International Symposium on Magnet Technology (MT-1)*. (Stanford, USA, 1965), pp. 431–446
5. R. Meinke, P. Schmüser, Y. Zhao, Methods of harmonic measurements in the superconducting HERA magnets and analysis of systematic errors, in *DESY report HERA* (1991), Report Number. 91–13
6. N. Smirnov, L. Bottura, F. Chiusano, O. Dunkel, P. Legrand, S. Schloss, P. Schnizer, P. Sievers, A system for series magnetic measurements of the LHC main quadrupoles. IEEE T. Appl. Supercon. **12**(1), 1688–1691 (2002)
7. P. Schnizer, *Measuring system qualification for LHC arc quadrupole magnets*. PhD thesis, TU Graz, 2002
8. J. Billan, L. Bottura, M. Buzio, G. D'Angelo, G. Deferne, O. Dunkel, P. Legrand, A. Rijllart, A. Siemko, P. Sievers, S. Schloss, L. Walckiers, Twin rotating coils for cold magnetic measurements of 15 m long LHC dipoles. IEEE T. Appl. Supercon. 1422–1426 (2000)
9. L. Walckiers, Z. Ang, J. Billan, L. Bottura, A. Siemko, P. Sievers, R. Wolf, Towards series measurements of the LHC superconducting dipole magnets, in *Proceedings of the Particle Accelerator Conference, 1997*. vol. 3 (May 1997), pp. 3377–3379
10. A.K. Jain, Harmonic coils, in *CAS Magnetic Measurement and Alignment*, ed. by S. Turner (CERN, August, 1998), pp. 175–217
11. W.G. Davies, The theory of the measurement of magnetic multipole fields with rotating coil magnetometers. Nucl. Instr. Meth. Phys. Res. Sect. A. Accelerators, Spectrometers, Detectors and Associated Equipment **311**(3), 399–436 (1992)
12. A. Devred, M. Traveria, Magnetic field and flux of particle accelerator magnets in complex formalism. *not published* (1994)
13. N.L. Smirnov, *Private Communication*
14. Sandwich quadripole n. 41. Coil calibration sheet for LHC quadrupole coil probe, (July 2002)
15. J. Billan, Calibration of the harmonic coil system for LHC magnet measurements, in *10th International Magnet Measurement Workshop (IMMW X)*, (Fermilab, Batavia, USA, Oct 1999)
16. O. Dunkel. Magnetic calibration system for rotating-coil-based measurement instrumentation at CERN, in *15th International Magnetic Measurement Workshop (IMMW XV)*, (Fermilab, Batavia, Illinois, USA, August 2007)

Chapter 6
Experimental Setup

The field descriptions outlined in the previous chapter show how fields in elliptical geometry can be described next to representations, which follow the curved trajectory of a charged particle in a dipole field. These developments reveal their real value if the coefficients of the associated multipoles are derived from measurements. These coefficients are then the basis of particle tracking studies and allow forecasting the performance of a machine which is still in the design or construction phase.

Measuring the field of superconducting accelerator magnets requires

- a test facility, providing a location and appropriate installation to connect the magnet to the cryoinfrastructure,
- a power converter to drive the magnet,
- a magnetic measurement system for measuring the field,
- and software infrastructure to control all the different systems required for setting up and controlling the different devices, taking the data and controlling the sequence of measurement.

This chapter is thus devoted to all the different systems required to make a test setup and its subcomponents, which are important to measure the magnetic field properties of SIS100 magnets.

6.1 Test Facility

The measurement data presented in the next chapters were taken at the Prototype Test Facility (PTF), which has been installed starting 2001 [1]. A photo of this facility is given in Fig. 6.1. The PTF consisted of the following components [1], when the measurements were made on which the results given here are based on:

© Springer International Publishing AG 2017
P. Schnizer, *Advanced Multipoles for Accelerator Magnets*, Springer Tracts
in Modern Physics 277, DOI 10.1007/978-3-319-65666-3_6

Fig. 6.1 The test facility for superconducting magnets at GSI. The C2LD magnet is currently connected to the feed box (*FB*) on the *left*. The universal cryostat is visible on the *right*. On the *left* the power converter (*PC*) currently being refurbished is shown

- a power converter to deliver 11 kA, 100 V with an accuracy of better than 10^{-4} and ramp rates well above 15 kA/s,
- a direct current current transformer (DCCT) and a digital volt meter HP3458A for measuring the current output of the power converter,

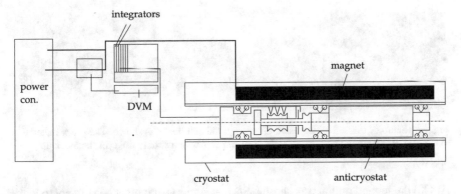

Fig. 6.2 Simplified setup showing the required systems for making the magnetic measurement as seen from the side. The magnet is housed by its cryostat with the anticryostat providing an room temperature access. The magnet is powered by a power converter. The current is transformed to an equivalent voltage which is measured with a high precision DVM. The magnetic measurement system "mole" is placed within the anticryostat. This mole contains slip rings. The output of the coil probes is connected to integrators. These are triggered by the angular encoder. At the end of the cryostat tables are mounted which allow moving the cryostat in the horizontal plane

- two test benches with their feed boxes and current leads,
- a cryoplant which could deliver 300 W at a temperature of 4.5 K and
- various instrumentation to monitor the thermodynamic state of the measurement object.

Further different auxiliary equipment is also available (vacuum pumps to obtain the vacuum cryostat, leak detection systems and so forth). This test facility was upgraded recently [2]. The power converter was scaled up for testing the SIS100 magnets, so that it can now deliver 20 kA, 22 or 66 V and ramp rates of 25 kA/s and above. The current leads were replaced with ones based on high temperature superconductor technology [3–5], as the cooling capacity would not have been sufficient to cool the magnet and conventional current leads. The test setup for magnetic measurements is sketched in Fig. 6.2.

6.2 The Anticryostat

The aperture of a superconducting magnet is typically exposed to the low temperature environment, which is required by the chosen superconductor. Thus two options to measure the field are available:

- operate the magnetic measurement equipment at cryogenic temperature or
- install an anticryostat, which insulates the warm inside from the cold surrounding of the magnet.

Fig. 6.3 Setup of the test anticryostat. On the *left* the end flange is shown. The *green* material is made of G10. The metal part on the *right* is an end cap, which is only used for the mock up

At GSI the second option was chosen. Such an anticryostat must be slim not to loose to much aperture and it must be of low thermal conductivity. These anticryostats are typically made of stainless steel [6, 7]. However the anticryostat fabricated for this test facility is foreseen for measurements of fast ramped superconducting magnets. Thus G10 (glass fibre reinforced epoxy) is used as material for the tubes. As the anticryostat is part of the cryostat its vacuum leakage must be compatible to the cryostat leak rate. Lacking sufficient experience with G10 cryostats, a separate insulation vacuum was foreseen, and thus two tubes are required. The wall thickness of these tubes is 2.5 mm and 3.5 mm respectively. These tubes are mounted concentrically with a gap of 2.5 mm in between. Superinsulation is wrapped between them. In that way a construction setup is achieved, which will not have eddy currents introduced into the tubes, but also the heat leak to the cold mass is limited. The total wall thickness of the anticryostat is 8.5 mm. (The test anticryostat is shown in Fig. 6.3). The wall thickness of the anticryostat reduces the available space within the inner tube to 47 mm. The outer diameter of the anticryostat is 64 mm thus being 4 mm less than the magnet aperture. The anticryostat is attached to two movable supports at both ends, which allow translating it laterally in the rectangular aperture of the magnet. The use of this feature is described in Sect. 8.1. Even if the magnet is only 3.1 m long, the total length of the anticryostat is ≈ 7 m, because of the diameter of the feed-box, the length of the cryostat feed-box connector and the length of the end cap.

Heat is constantly flowing from the anticryostat to the cold mass. Therefore heaters are installed on the outer surface of the inner tube of the anticryostat, which are used to stabilise its internal temperature. These are electrically powered and could disturb the magnetic field. The heaters consist of 4 copper wire pairs, which are wrapped around as helices similar to a screw threading with 4 starts. These pairs are electrically connected at the end to form a bifilar winding. Three of these windings are used as heaters while the resistance of the remaining one of them is used to measure the mean temperature of the anticryostat. A simple parameter regulator is used to control the temperature. Its analogue set value is used to steer a power converter whose voltage is controlled by a reference value. This total setup was made to minimise the magnetic field distortions induced by the electric heating system. The total heat loss amounts to ≈ 7 W for the anticryostat.

6.3   Magnetic Measurement Equipment

6.3.1   History: Choice of Method

Magnetic measurement equipment for measuring accelerator magnets is typically constructed and fabricated at the lab, which is responsible for the whole project or the accelerator magnet construction, especially if superconducting magnets are involved. A comprehensive overview on magnetic measurement methods is given in [8, 9] and some chapters in [10].

The magnet's aperture is material free and thus the field can be described by the MQS theory of linear material (see Sect. 2.1.2). Hence a scalar potential fulfilling Laplace's equation can be used. It is one of its properties that the values on the boundary of the domain will be extremal. Experience has shown that extrapolation is cumbersome, so accurate measurements require to cover as much of the area of interest as possible. This will avoid extrapolation and the measurement range of the chosen instrument will be utilised to a large extend. Accelerator operation requires that the following properties of the main magnets are well known:

Integral dipole field   In accelerators all main magnets are connected in series, where each dipole shall deflect the beam by the same amount. A variation in its deflection will lead to deviations from the ideal orbit.

Integral dipole angle   Any deviation will bend the particle off the plane of the orbit.

Integral quadrupole field   The quadrupole lenses correct particle paths which are off orbit. An deviation will cause that these corrections will create artefacts on these corrections.

Integral quadrupole axis   A quadrupole which is off axis will introduce a kick each time the particles passes by. In a circular machine this kick occurs at every revolution when the particle passes by and will increase the transverse movement turn by turn. This movement can get so large that the particle hits the aperture and the particle is lost eventually.

The field deterioration of the magnet ends is dominated by other effects as the field inhomogeneity in the magnet centre: e.g. for iron dominated magnets the field quality in the magnet centre is dominated by the lamination geometry while at the end it is dominated by the coil positions. Therefore measuring only the total field integral is typically not sufficient. The integral field has to be measured separately for the magnet end and the central part of the magnet. Coil probes, measuring the flux and integrating over a certain area (see Chap. 5), are a natural choice, as they allow measuring the integral over a certain length. This length can be chosen within the constrains defined by mechanical fabrication, component availability, signal strength and so on. Furthermore harmonics (i.e. cylindrical circular multipoles) can be derived from the measurement results of rotating coils (see Sect. 5.3).

Measurement systems, based on rotating coil probes, have proven to be able to measure reliably integral circular harmonics within the area covered by them. Besides the coil probe the following components are required for a full system:

- ball bearings,
- angular encoder,
- drive (typically a motor),
- inclinometer to relate the angle to gravity.

As the measurement system must not distort the object investigated, magnetic measurement systems must be transparent to the magnetic field, thus $\mu_r = 1$ must be fulfilled with sufficient precision. Furthermore any moving part must be made of material of high electric resistivity. Otherwise eddy currents are induced in these parts and thus magnetic fields. These fields can deteriorate the to be measured field or large forces have to be applied on the object to move it, as the forces caused by the interaction with the magnetic field interaction have to be overcome, which can result in an inacceptable deformation of the measurement device.

The different system components listed above are not necessarily exposed to magnetic fields, nor is the field distortion they create negligible. Precision ball bearings are typically made of steel, angular encoders can be made of metal strips and inclinometers can be based on a pendulum with a magnet as part of an induction break to dump the pendulums oscillations.

Measurement systems using rotating coil probes have been built in different styles:

Free hanging   In a "free hanging" system the coil probe is hanging freely within the magnet aperture and is not supported within the magnet aperture. The coil probe is long enough that its ends are not exposed to the magnetic field any more. Thus, apart from the coil probe, all parts can be chosen from standard industrial components. Examples are e.g. [11, 12].

Shaft   Shaft based systems use a mechanical rigid connection (a shaft) between the coil probe and all auxiliary systems which are placed outside of the magnetic field. The ball bearings, controlling the coil probes rotation, are mounted closely on the coil probe, and thus are exposed to the magnetic field. For these systems only the coil probe, the shaft and the required bellows and ball bearings have to be made of high resistance and magnetic transparent material. Examples of such systems are given in [13–15].

Mole   A system is called a "mole", when all the auxiliary components are operated within the magnetic field. This reduces the physical principles they may be based on. The reduced set may be further narrowed if a component supplier uses parts which are made of magnetic material or of too low electrical resistance (if moving or exposed to fast ramped magnetic fields). The advantage of these systems is that they are small and self contained. If the magnet aperture is small compared to the length of the magnet or the magnet measurement bench, the smaller mechanical lengths allow meeting the requirements of the field angle measurement [16–18]. Typically the magnet test bench setup is considerably longer for a superconducting magnet, as commonly the magnet cryostat has to be closed by a end cap on one side and an interconnection piece and a feed box on the other side.

While the magnet length of SIS100 is modest (dipole 3 m), the end cap of the cryostat and the length of the feed-box give a total length of 7 m. The aperture itself is only 47 mm within the anticryostat; thus a mole approach is favourable. Furthermore the selection of a mole style system was made, when a few of the projected FAIR machines were utilising superconducting magnets of similar aperture (SIS100, SIS300 and HESR).

Moles have been built for different machines before:

SSC   Here an air motor was used to generate the drive. An air motor requires high rotation speed while rotating coils are typically used at low speed to minimise artefacts due to vibrations. It required a gear box and a damper to control the oscillations [16].

LHC   The magnetic field of all LHC main magnets was measured at the producers as part of production using a mole [17]. This approach allowed only measuring the magnetic field of a subsample of the magnets during the cold testing campaign.

Both these moles were made to measure stationary fields. The high ramp rates of SIS100 make it, however, even more challenging as all materials used must be non conducting and non magnetic. Furthermore all cabling has to avoid flux loops in any particular place.

6.3.2   A Modular Mole

When the modular mole was projected, it was targeted for measuring the field of magnets of three different machines: SIS100, SIS300 and HESR. Therefore the system was designed so that many of its components can be easily reused. Thus the mole is split up in individual pieces which can be adapted to the different diameters of these machines and then assembled as required (see Figs. 6.4 and 6.5).

The motor unit consists of a piezo motor rotating the coil probe and auxiliary equipment. The shaft connecting the piezo motor to the coil probe is hollow, so that the cables, carrying the coil probe signals can be placed inside the shaft. Slip rings are then used to tap the signals to the non rotating body of the mole. An encoder with 512 counts is mounted on the shaft as close as possible to the coil probe, which allows triggering the integrators when measuring static magnetic fields. The field coil probe is connected to the plug, to which an axis coil probe can be connected. As no commercial high resolution encoder with a hollow axis was found fitting into the aperture of the anticryostat (inner diameter 47 mm) a separate unit with a 18 mm diameter optical encoder with 7500 counts was built, which can be mounted on the other side of the coil probe array. The encoder has two quadrature channels (i.e. they are 90° phase shifted). Thus its counts can be quadrupled and the coil probe angle can be measured with a precision of 0.4 mrad. An additional stationary coil probe is mounted in the motor unit, which allows verifying the field reproducibility from ramp to ramp. All angular measurements are referred to gravity using electrolytic inclinometers. Two of them are mounted inside the motor unit, with

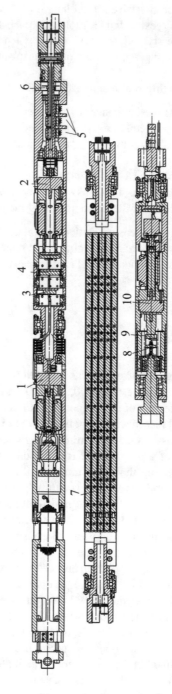

Fig. 6.4 The different modules of the mole. The motor unit is shown on *top*, the field coil probe and the precision encoder unit follow below. The leveling piezo motor...*1*, the coil rotation piezo motor...*2*, the inclinometers...*3* and *4*, the slip rings...*5*, the angular encoder with 512 counts...*6*, the coil probes...*7*, the angular encoder with 7500 counts...*8*, its inclinometer...*9* and leveling motor...*10*

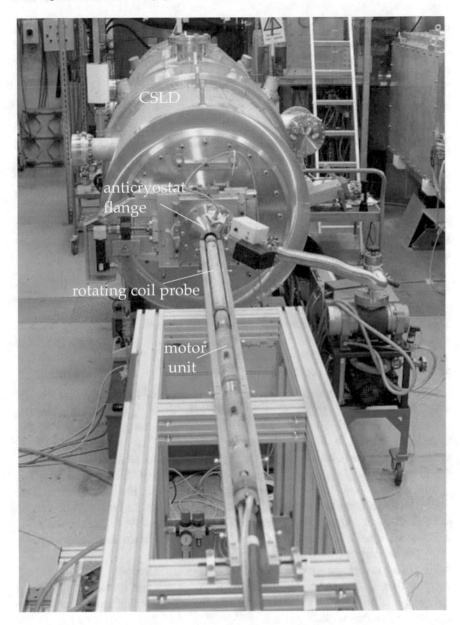

Fig. 6.5 The mole in front of the SIS100 FoS dipole magnet (CSLD), ready for installation in the magnet. The anticryostat is installed within the magnet. Only the end flange with its connection to the vacuum pump is visible. The rotating coil is next to the magnet and the motor unit together with the auxiliary unit close to the vertical end. Photo courtesy of G. Otto, GSI Helmholtzzentrum für Schwerionenforschung mbH

an angle of 45° between them, as these inclinometers can give fake signals in the normal measurement range if they are turned up side down. In this case the second inclinometer will show saturation on the wrong side. This is used by the motion controller to turn them by 180° to the proper measurement range.

6.3.2.1 General Solutions

The housing of the different systems is made of G11. All ball bearings are made of ceramic and are machined to high precision. Slip rings are used to get the signals of the coil. The inclinometers are of the electrolytic type. These were tested in a dipole up to two Tesla and found to be sufficiently accurate. The bellow coupling the motor shaft to the coil and the rollers with its springs are metallic. The bellows are made of an TiVaAl alloy, which is non magnetic. The rollers are made of brass and the springs on the rollers are made of CuBe following the design of the rollers used at CERN for measuring the LHC magnets.

6.3.2.2 Coil Probe

The coil probe used for all the measurement data given in the following chapters was a coil probe originally developed for the LHC project [14]. It consists of 5 radial coils. The coil probe and its properties are described in Chap. 5. Its fabrication is made the following way (developed by the late Jacques Billan and his team):

- Individual coils are fabricated. Around 30–50 coils have to be fabricated.
- Each coil is turned upside down in a reference dipole and the flux change is measured. From the induced flux change the area of the coil probe is deduced using the relation $\Phi = \int \vec{B} \, d\vec{\sigma}$.
- The coils are sorted for their area and pairs of similar area are identified. The pair with the most similar area form the coil probe A and C; the following pair is selected for B and D. (see also Fig. 5.4 page 82). As coil E a coil is chosen from the remaining ones, whose area matches best the area of the coils A—D.
- The individual coils are then assembled in the middle of the support structure.
- The whole assembly consisting of the coils and the support structure is remachined, to obtain a perfect cylinder.
- Finally the ball bearing pieces are mounted onto the assembly.
- Only when all the steps above have been completed and the whole coil probe assembly is finished the rotation radii of the individual probes can be measured in a reference quadrupole.

6.3.2.3 Angular Positioning

This modular mole is built for measuring static and transient magnetic fields. The measurement of transient harmonics is based on the "step-by-step method" [19–24]. This method uses the following procedure:

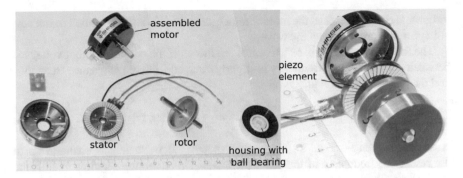

Fig. 6.6 The Shinsei piezomotor and its parts

1. the coil probe is positioned to the start angle,
2. the magnet is ramped in transient mode and the field change is measured,
3. the coil probe is rotated by some incremental angle,
4. then steps 2 and 3 are repeated until the full circle is finished.

After these steps the magnetic field is computed for each moment in time in question and then the harmonics are computed.

The step by step method requires that the coil probe can be precisely positioned in angle [24]. The only piezomotor with significant momentum commercially available (already utilised at CERN for an other mole [18]) at this time had a controller, which did not allow controlling the motor rotation to slow speeds. Given that piezomotors are frequently used as positioners, a study was conducted [24, 25], which showed that a precise angular positioning can be reached (better than 0.1 mrad). The piezomotor travelling wave is generated by two electrical wave sent to the piezo elements of the motor, (see Fig. 6.6 for a setup of the motor) which make the elements on the stator form a travelling wave, pulling the rotor forward. The controller delivered together with the piezo motor was limited in the angular accuracy it could achieve as it always keeps the same angle within the electrical waves at an angle of $\alpha = \pi/2$. Furthermore the motor motion is based on a resonance effect. This effect is thus sensitive if the exciting wave matches the eigen frequency. The controller, delivered with the motor, uses one of the stator elements as reference element to measure the excitation and tunes the frequency using a PLL to the amplitude measured. Hence a minimum excitation of the stator elements is required so that enough movement is induced in the reference element so that the PLL can operate.

As a consequence a new controller was built, which is controlled by an applied voltage. For maximum speed the motor is excited with the two waves

$$w_1 = A \sin(\omega_r t) \quad \text{and} \quad w_2 = A \sin(\omega_r t + \pi/2), \tag{6.1}$$

where A is the amplitude (roughly 150 V) and ω_r the resonance frequency ($\approx$51.5 kHz). The resonance frequency was measured on a test bench next to the torque and rotational speed the motor provides as a function of the applied frequency, voltage and phase between the waves.

The following mode is used to control the motor. If the controller input voltage is between -5 and 5 V, the frequency is slightly detuned from the resonance frequency and the phase α between the waves

$$w_1 = A \sin(\omega_r t) \quad \text{and} \quad w_2 = A \sin(\omega_r t + \alpha) \tag{6.2}$$

is tuned between $-\pi$ and π. For higher speeds the frequency is changed up to the resonance frequency of the piezo elements. This detuning requires to shift the frequency with an accuracy of ≈ 10 Hz. Piezo motors must not be activated constantly as these are based on friction. In particular at 0 V, the phase between the waves is 0. Both groups of elements are still excited but do not provide any movement. Therefore an additional input was provided which allows switching the whole amplifier on or off. This whole signal generation is realised in a Field Programmable Gate Array (FPGA) using numerically controlled oscillators (see Fig. 6.7). The voltages are then produced by digital to analog converters (DAC), whose output is only ± 10 V and only little current can be drained and thus requires to be amplified. The piezomotors are capacitive devices; the original controller's amplifier size is minimised using transformers with a inductance, which matches the capacitance of the motor. The capacitance of the connecting cable is comparable to a significant fraction of the motor capacitance and thus can not be neglected. So the signal amplifier was rebuilt using a strong amplifier (see Fig. 6.8) together with a medium frequency transformer and a matching inductance so that a broad band resonance is formed [24, 26]. In this case the precise cable length does not matter. The motor can be controlled in a

Fig. 6.7 The control loop of the piezo motor. The frequency generator is realised in a field programmable array. DAC...digital analog converter, ADC...analog digital converter

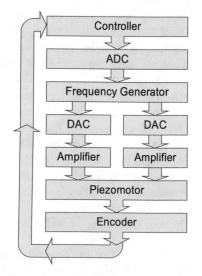

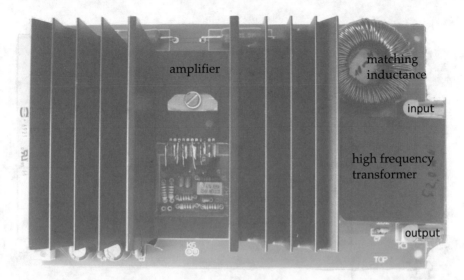

Fig. 6.8 The power driver for the piezomotor

closed loop down to an angle of 0.4 mrad only limited by the angular encoder counts. Further details on the mole development can be found in [27].

6.3.2.4 Data Acquisition

The signal from the coils has to be processed and analysed to yield the field properties. (The mathematical discussion was given in Sect. 5.3). The first signal processing steps in these chain are performed by analogue and digital electronics.

To get the compensated signal the coils are galvanically connected on the patch panel as shown in Fig. 6.9. The output of both systems is then amplified by preamplifiers. To obtain better independence from rotation instability the signal is integrated [28]. The angular encoder triggers the readout of the integrator and ensures that the flux is measured at equally spaced angular steps. The integrators [29, 30] are fabricated as VME Cards. An external controller then transfers the data to a computer for online measurement control and further data analysis.

The integrator consists of two main components:

1. A voltage to frequency converter (VFC). A voltage of 5 V is converted to a frequency of 250 kHz. With decreasing voltage the frequency decreases as well. A voltage of −5 V Volts is converted to 0 Hz.
2. A counter. It counts the pulses coming from the VFC. The total counts are equal to the input of the VFC integrated over time.

The integrator is shipped with additional logic supporting a triggered readout of the counter and shift registers in order not to loose any input during readout.

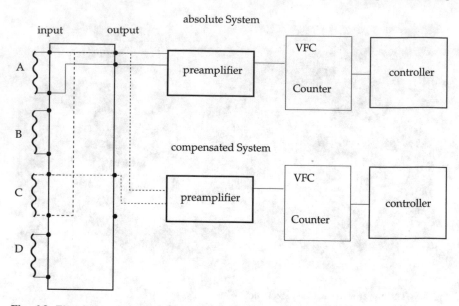

Fig. 6.9 Electrical system setup. The *solid lines* indicate the cabling of the absolute coil. The *dashed lines* indicate the signal cables of the compensated system. The cables are connected to a preamplifier. The Voltage to Frequency Converter (*VFC*) and the counter form the integrator. A controller reads the signal out

6.3.2.5   Measurement Accuracy

The mole's measurement accuracy was found to provide

- an accuracy of the dipole harmonic C_1 of 5×10^{-4} comparing the measurements of the mole to a hall probe,
- and an random error of less than 0.1 units for the higher order harmonics.

The measurement of the field direction versus gravity is still under evaluation. This value is not required for the results presented in this treatise.

6.4   Summary

Testing superconducting magnets and measuring their fields requires a significant amount of infrastructure. It consists of a cryoplant, a power converter and safety systems.

An anticryostat built of G10 tubes allows accessing the cold bore at room temperature. A mole was built and used to measure the first SIS100 model dipole magnets

(S2LD, C2LD) and the first of series dipole magnet (CSLD). The coil is rotated by a piezo motor with a dedicated controller built, which allows rotating the motor even at very small speeds.

References

1. A. Stafiniak, E. Floch, P. Hahne, G. Hess, M. Kauschke, F. Klos, F. Marzouki, G. Moritz, H. Mueller, M. Rebscher, P. Schnizer, C. Schroeder, G. Walter, F. Walter, H. Welker, Commissioning of the prototype test facility for rapidly-cycling superconducting magnets for FAIR. IEEE Trans. Appl. Supercond. **18**(2), 1625–1628 (2008)
2. P. Schnizer, A. Mierau, A. Bleile, V. Maroussov, A. Stafiniak, W. Freisleben, H. Raach, J.P. Meier, Low temperature test capabilities for the superconducting magnets of FAIR. IEEE Trans. Appl. Supercon., **25**(3), **6**(20150). Art.Nr: 9500505
3. A. Ballarino, A. Ijspeert, Design and test of the prototype high t/sub c/ current leads for the large hadron collider orbit correctors. IEEE Trans. Appl. Supercon. **5**(2), 805–808 (1995)
4. A. Ballarino, Current leads for the LHC magnet system. IEEE Trans. Appl. Supercon. **12**(1), 1275–1280 (2002)
5. H. Raach, C. Schroeder, E. Floch, A. Bleile, P. Schnizer, T. Andersen, 14 kA HTS current leads with one 4.8 K helium stream for the prototype test facility at GSI, in *Physics Periodica*, (2015 Vol. 67, pp. 1098–1101)
6. F. Clari, O. Dunkel, M. Genet, C. Gregory, P. Sievers, Prototype development of a warm bore insert for the LHC magnet measurements. IEEE Trans. Magn. **30**(4), 2662–2664 (1994)
7. O. Dunkel, P. Legrand, P. Sievers, A warm bore anticryostat for series magnetic measurements of LHC superconducting dipole and short straight section magnets. AIP Conf. Proc., **710**(494–504), 9 (Jan 2004)
8. CERN, *CAS–CERN Accelerator School: Magnetic measurement and alignment*, (CERN, Geneva, 1992)
9. CERN, *CAS–CERN Accelerator School: Measurement and Alignment of Accelerator and Detector Magnets*, (CERN, Geneva, 1998)
10. CERN, *CAS–CERN Accelerator School: Specialised course on Magnets*, (CERN, Geneva, 2010). Comments: 14 lectures, 494 pages, published as CERN Yellow Report, http://cdsweb.cern.ch/record/1158462
11. O. Pagano, P. Rohmig, L. Walckiers, C. Wyss, A highly automated measuring system for the LEP magnetic lenses. J. Phys. Colloques, **45**(C1):C1–949–C1–952, (1984)
12. L. Walckiers, Magnetic measurements on the ISR superconducting quadrupoles, in *IEEE Transactions on Magnetics*, **17**(5):1872–1874, (Sep 1981)
13. J. Billan, J. Buckley, R. Saban, P. Sievers, L. Walckiers, Design and test of the benches for the magnetic measurement of the lhc dipoles. IEEE Trans. Magn. **30**(4), 2658–2661 (1994)
14. N. Smirnov, L. Bottura, F. Chiusano, O. Dunkel, P. Legrand, S. Schloss, P. Schnizer, P. Sievers, A system for series magnetic measurements of the LHC main quadrupoles. IEEE Trans. Appl. Supercon. **12**(1), 1688–1691 (2002)
15. L. Walckiers, Z. Ang, J. Billan, L. Bottura, A. Siemko, P. Sievers, R. Wolf, Towards series measurements of the LHC superconducting dipole magnets, in *Proceedings of the Particle Accelerator Conference, 1997*, vol. 3 (May 1997), pp. 3377–3379
16. G. Ganetis, J. Herrera, R. Hogue, J. Skaritka, P. Wanderer, E. Willen, Field measuring probe for SSC magnets, in *Proceedings of the 1987 IEEE Particle Accelerator Conference* (1987), p. 1393
17. J. Billan, S. De Panfilis, D. Giloteaux, O. Pagano, Ambient temperature field measuring system for LHC superconducting dipoles. IEEE Trans. Magn. **32**(4), 3073–3076 (1996)

18. L. Bottura, M. Buzio, G. Deferne, C. Glöckner, H. Jansen, A. Köster, P. Legrand, A. Rijllart, P. Sievers, A mole for warm magnetic and optical measurements of LHC dipoles. IEEE Trans. Appl. Superconduct. **1**, 1454–7 (2000)

19. K. Gertsev, Y. Luzhin, N. Smirnov, P. Scherbakov, S. Trofimov, A. Zlobin, The research of the harmonics time dependence of the UNK SC-dipole magnets, in *Proceeding of XI ECFA*, (1990)

20. K. Gertsev, Y. Luzhin, N. Smirnov, P. Scherbakov, S. Trofimov, A. Zlobin, Dynamic effects in the magnetic field of the SC-dipole for the UNK. Preprint, (1987)

21. K. Gertsev, Y. Luzhin, N. Smirnov, S. Trofimov, Measurement of the dynamic additions to the magnetic field in the UNK. Preprint, (1987)

22. A.D. Kovalenko, *Private communication*

23. A. Daël, *Private communication* (2006)

24. P. Schnizer, H.R. Kiesewetter, T. Mack, T. Knapp, F. Klos, M. Manderla, S. Rauch, M. Schöncker, R. Werkmann, Mole for measuring pulsed superconducting magnets. IEEE Trans. Appl. Supercon. **18**(2), 1648–1651 (2008)

25. Maximilan Manderla, Positionsregelung mit einem Ultraschall-Wanderwellen-Motor, Modellbildung und Regelung in Simulation und Experiment. Master's thesis, (TU-Darmstadt, Germany, 2007)

26. R. Werkmann, *Leistungselektronik zum Ansteuern von 3 Piezomotoren* (Technical report, GSI Helmholtzzentrum für Schwerionenforschung mbH, 2006)

27. Thomas Mack, *Der Maulwurf–Entwicklung und Fertigung eines komplexen Magnetmesssystems* (Diplomarbeit Hochschule Darmstadt, Juni, 2008)

28. A.K. Jain, Harmonic coils, in *CAS Magnetic Measurement and Alignment*, ed. by S. Turner (CERN, August 1998), pp. 175–217

29. G. Turcato, *Desription du module electronique "integrateur 32 bits binaires programmable" compatible au bus G64* (CERN, january, Note interne, 1983)

30. P. Galbraith, Portable digital integrator (Technical report, CERN, 1993)

Chapter 7
Applications

7.1 Appropriate Handling of Calculation Data

Field data obtained by numerical codes will be affected by numerical errors or limits given by the accuracy of the algorithm. These data are commonly available on a grid and not necessarily on the curve required for calculating the coefficients of the multipoles. So an interpolation is required which can use the fact that the field data are calculated for accelerator magnets. These accelerator magnet fields are dominated by one circular multipole, the main multipole $\mathbf{C_m}$. All further field deviation is small; thus the total interpolation quality is improved if only the deviation of the field from the ideal multipole field produced by $\mathbf{C_m}$ is interpolated. A wrong choice for $\mathbf{C_m}$, in particular of m, will create additional artifacts.

A useful practical interpolation of a magnetic field $\mathcal{B}(x, y)$ of an accelerator can be obtained by:

1. In a first step one interpolates the field by bilinear interpolation. Two interpolation variables λ and μ are defined by

$$\lambda = \frac{x - x_i}{x_{i+1} - x_i} \quad x_i \leq x \leq x_{i+1}, \quad \text{and} \quad \mu = \frac{y - y_i}{y_{i+1} - y_i} \quad y_i \leq y \leq y_{i+1}.$$
(7.1)

The field $\mathcal{B}(x, y)$ is interpolated using bilinear interpolation

$$\mathcal{B}(x, y) = (1 - \mu) \left[(1 - \lambda) B_g(x_i, y_i) + \lambda B_g(x_{i+1}, y_i) \right] + \quad (7.2)$$
$$+ \mu \left[(1 - \lambda) B_g(x_i, y_{i+1}) + \lambda B_g(x_{i+1}, y_{i+1}) \right]$$

with B_g the values at the different grid points. The interpolation above is still plagued by the limitations of linear interpolation and thus will not yet yield the required quality of the interpolation. Simply consider a sextupole. The field value depends on $\mathbf{z}^2$ thus linear interpolation will only give limited accuracy. But for accelerator magnets the main component is the dominant one, while the other components are smaller by orders of magnitude.

© Springer International Publishing AG 2017
P. Schnizer, *Advanced Multipoles for Accelerator Magnets*, Springer Tracts
in Modern Physics 277, DOI 10.1007/978-3-319-65666-3_7

2. The component of the main multipole C_m is obtained by a suitable fit. For the dipole value one simply uses the mean value along the reference curve. For the quadrupole a linear fit like $B_y = gx$ will yield the desired value. Any reference curve can be used, as long as it is adapted to the domain of the problem the data are used later on.
3. The field given by the main multipole is now used to calculate its contribution at each point z_i

$$B_i(z_i) = C_m \left(\frac{z_i}{R_{Ref}} \right)^{n-1}$$ (7.3)

$$B_i(x_i, y_i) = \begin{pmatrix} \text{Re}(B_i) \\ \text{Im}(B_i) \end{pmatrix}$$

with $z_i = x_i + \iota y_i$. The B_i are subtracted from the grid data to obtain only the contribution of the other multipoles to the field data. This correction field B_c to the main multipole is now given by

$$B_c(x_i, y_i) = B_g(x_i, y_i) - B_i(x_i, y_i)$$ (7.4)

Given that the deviation from the main field of an accelerator magnet is below 0.1%, the approximation gains 2–3 digits without the loss of any generalisation.
4. Now the interpolated field is calculated using

$$\mathcal{B}(x, y) = B_i(x_i, y_i) + (1 - \mu) \left[(1 - \lambda)B_c(x_i, y_i) + \lambda B_c(x_{i+1}, y_i) \right]$$ (7.5)
$$+ \quad \mu \quad \left[(1 - \lambda)B_c(x_i, y_{i+1}) + \lambda B_c(x_{i+1}, y_{i+1}) \right] .$$

Steps 2–4 can be repeated, reevaluating the field on the reference curve to obtain a better approximation for the main multipole. This approach has proven to provide a better interpolation than a pure linear approach. The multipole coefficients for quadrupole magnet fields showed a better representation of the original field when this algorithm was used in comparison to a pure linear interpolation. More than three iterations did not prove to be useful. If multipoles have to be calculated from grid data the algorithm presented here is used for interpolating the data along the reference curve.

7.2   Calculating Cylindrical Elliptical Multipoles

The formulae described in the previous chapters (see Sect. 4.2) are now applied to dipole fields to demonstrate that all these steps are necessary to interpolate the field within an ellipse with a precision of better than the maximum tolerable field deviation of 600 ppm or 6 units (1 unit corresponds to 100 ppm). Mathematically speaking the field quality $\Delta \mathbf{B}$ of a dipole (in units) is given by

$$\Delta \mathbf{B}(\mathbf{z}) = \frac{\mathbf{B}(\mathbf{z}) - \mathbf{B}(\mathbf{0})}{\mathbf{B}(\mathbf{0})} \, 10^4 . \tag{7.6}$$

The higher order harmonics (for the dipole) are given by

$$b_n + \mathrm{i} a_n = \mathbf{c_n} = \frac{\mathbf{C_n}}{\mathbf{C_1}} \, 10^4 . \tag{7.7}$$

The field quality is calculated for the **C**urved **S**ingle **L**ayer **D**ipole (CSLD), [1–3], the dipole design chosen for the main dipole for the SIS100 machine of FAIR. The original distribution of its aperture field is given in Fig. 7.1a at a current of 873 A yielding a field of $\approx$0.13 T. The field was taken along the elliptical boundary of the good field region ($a = 57.5$ mm and $b = 30$ mm) and the elliptical multipoles are calculated as defined in (4.29). Using the first 20 coefficients the field was interpolated within the aperture (see Fig. 7.1b). The naked eye can not see any difference to the original data (Fig. 7.1a). The original field was subtracted from the interpolated one. One can see from Fig. 7.1c that this difference is well below half a unit and thus the representation based on the elliptical coefficients is sufficiently precise that beam dynamics can evaluate the distortions created by this magnet. Beam dynamic codes or end users expect circular multipole. Cylindric circular multipoles were calculated using a Fourier Transform of the data along a circle. Again the interpolation data were calculated (see Fig. 7.1d) and its difference to the original data (see Fig. 7.1e) using the first 15 coefficients. One can see that the interpolation works well within the circle but outside the circle soon the errors get unacceptably large. The difference outside of the circle is even larger if more coefficients are used. At last the circular multipoles were calculated from the elliptical ones as described in 4.35 (see Fig. 7.1f for the interpolation and Fig. 7.1g for the difference). One can see that, contrary to Fig. 7.1d, e the interpolation works even outside the circle and within the whole ellipse. The same result was obtained for magnetic fields calculated for different SIS100 dipole and quadrupole configurations [4].

The plots of Fig. 7.1 show that the interpolation based on the elliptical representation achieves an interpolation with sufficient accuracy; thus it is interesting to see how large the difference is. This is illustrated for the allowed circular harmonics of the aforementioned dipole along the load line (i.e. at different current points within the typical operation limits) in Fig. 7.2.

The multipoles were obtained once by calculating the elliptical multipoles (4.29) and converting them to circular multipoles (Sect. 4.2.2) and secondly by calculating the multipoles on the circle directly. The reference radius used was $R_{\mathrm{Ref}} = 40$ mm. The largest difference is found for the sextupole b_3 and is in the order of 0.2 units (= 20 ppm). For all higher multipoles it is not visible in this scale. Thus the difference is shown for all of them in Fig. 7.3.

Except for the sextupole the difference is less then 1 ppm! While these differences are small, these determine if the approximation is valid up to the border of the ellipse (see Fig. 7.1d, f). This shows that the representation is very sensitive to the size of its coefficients if cylindrical circular multipoles are used. The cylindrical elliptical ones

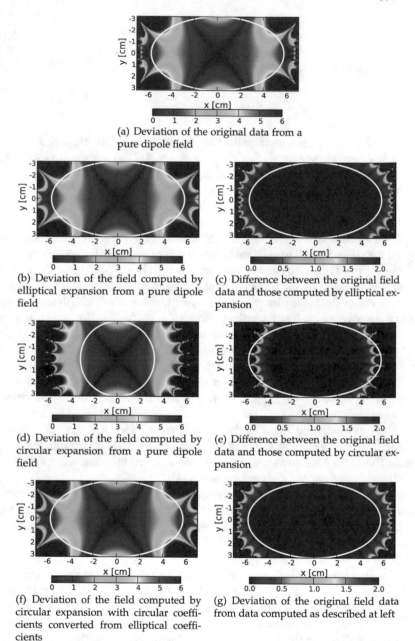

(a) Deviation of the original data from a pure dipole field

(b) Deviation of the field computed by elliptical expansion from a pure dipole field

(c) Difference between the original field data and those computed by elliptical expansion

(d) Deviation of the field computed by circular expansion from a pure dipole field

(e) Difference between the original field data and those computed by circular expansion

(f) Deviation of the field computed by circular expansion with circular coefficients converted from elliptical coefficients

(g) Deviation of the original field data from data computed as described at left

Fig. 7.1 Test of the field representation quality of the cylindrical circular and the cylindrical elliptical multipoles using simulation data obtained by FEM for the CSLD at a current of 873 A and a field of 0.13 T. The field B_y in the aperture is plotted. The *colour* indicates the absolute value of the deviation (in units). The original data are given on top. The *left column* shows the fields as obtained from the coefficients and the *right columns* shows the absolute value of the difference between the fields obtained from the coefficients and the original data.

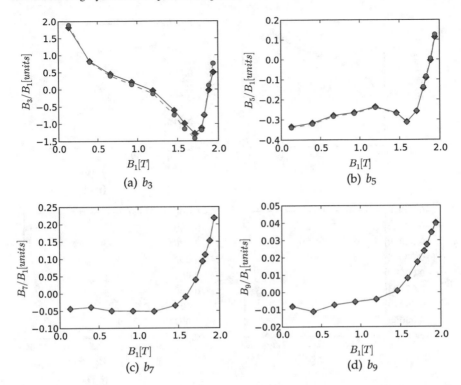

Fig. 7.2 The field quality for the 8 turn curved single layer dipole (CSLD) The *solid line* with the diamonds represents the circular multipoles as obtained from the elliptical multipoles calculated on the ellipse and the *dashed line* with the circular multipoles represent the circular multipoles as calculated on a circle. Please note that the changes are small but significant (e.g. sextupole, difference in the order of 0.2 units)

are not affected by this artifact, given that the geometry of the cylindrical elliptical coordinates is more adapted to the problem.

7.3 Summary

Accelerator magnets provide fields of good field homogeneity. Thus any interpolation required for further processing must be careful not to introduce artefacts of its own. Experience shows that one has to take care of this fact if multipoles or harmonics are to be calculated.

Cylindrical elliptical multipoles are used for describing the field of various magnets and it was demonstrated that these represent the field with sufficient accuracy and within a larger area than the circular multipoles calculated on a circle.

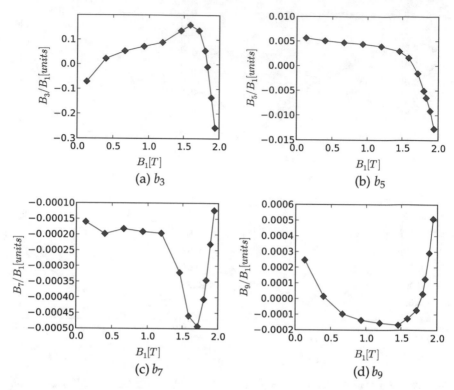

Fig. 7.3 The difference between the coefficient sets for the curved single layer dipole version (CSLD). The circular ones calculated from the data on the circle were subtracted from the circular ones obtained from the elliptical ones

This is explained by the fact that reconstructing the field outside of the circle, gets more and more problematic the higher order multipoles have to be taken into account, simply due to the fact that $(\mathbf{z}/R_{\text{Ref}})^n$ increases with n, if $|\mathbf{z}| > R_{\text{Ref}}$.

References

1. P. Akishin, E. Fischer, P. Schnizer, A single layer dipole for SIS100. Technical report, Gesellschaft für Schwerionenforschung mbH, Planckstraße 1, D-64291 Darmstadt, July 2007
2. P. Schnizer, B. Schnizer, P. Akishin, E. Fischer, Magnetic field analysis for superferric accelerator magnets using elliptic multipoles and its advantages. IEEE T. Appl. Supercon. **18**(2), 1605–1608 (June 2008)
3. FAIR—Facility for Antiprotons and Ion Research, Technical design report, Synchrotron SIS100, Dec 2008
4. P. Akishin, E. Fischer, P. Schnizer, Field quality of various designs of the SIS100 magnets. Technical report, Gesellschaft für Schwerionenforschung mbH, Planckstraße 1, D-64291 Darmstadt, July 2007

Chapter 8
Measuring Advanced Multipoles

The value of advanced multipole descriptions is only of practical use for accelerator and magnet designers if their coefficients can be deduced from measurement results. The aperture of iron dominated dipole magnets is typically rectangular and the vacuum chamber of elliptical shape. The standard approach for describing accelerator magnet fields are cylindrical circular multipoles. The coefficients are obtained from the field measured on the circular boundary. Within the rectangular magnet shape only a circle, which is significantly smaller than the ellipse, can be inscribed, so the cylindrical circular multipole coefficients have to be used to estimate the field beyond this circle to cover the whole aperture available to the beam. Such estimation beyond the original domain of the measurement data is typically difficult and error-prone thus advanced methods are necessary which allow describing the field within an ellipse. Rotating coil probes have proven as useful tool to derive multipoles. The path of a charged particle is curved within a dipole field and iron dominated magnets are frequently built as curved magnets. Toroidal circular multipoles are used to show that rotating coil probes can be used to measure curved magnets, if the curvature is not too large. Further limits of this approach are presented.

The measurement of these multipoles is exemplified on the data obtained on the first model magnet S2LD and on the CSLD, the SIS100 first of series dipole magnet (see Sect. 1.3). The magnetic field of the S2LD magnet can be described concisely with cylindrical elliptical multipoles. The CSLD ones require multipoles reflecting the curvature of the magnet; it will be shown, however, that cylindrical elliptical multipoles can be used to treat the fields of these curved magnets too, if appropriate correction actions are taken into account.

© Springer International Publishing AG 2017
P. Schnizer, *Advanced Multipoles for Accelerator Magnets*, Springer Tracts
in Modern Physics 277, DOI 10.1007/978-3-319-65666-3_8

8.1   Measuring Straight Elliptical Multipoles

8.1.1   Calculation Procedure

The magnetic field of the first full size SIS 100 dipole model magnet (S2LD) [1–4] is measured using a rotating coil probe [5], equipped with bucking coil probes (see Sect. 5.4). The same measurement procedure is applied to the first full size SIS100 curved single layer magnet (CSLD) [6, 7].

One could consider to measure the field along the elliptical boundary and deduce harmonics from these measurements. Such measurements, however, are impractical within an anticryostat, which is only of limited use as a mechanical reference. Furthermore a magnetic field probe would have to be moved along the ellipse with an accuracy of several μm. Therefore a different approach is chosen here: The field is measured at different lateral positions using a rotating coil probe and the measurements are then combined. The description given here was first presented in [8] with more explanations given in [9]. The text here builds on these two publications.

Below the method of combining the measurements is outlined based on the data taken from the central field or end field of theses magnets, as was considered most suitable for demonstrating the method. Finally the multipoles obtained with the method presented here are validated comparing the field calculated from theses multipoles to the field data measured along 3 lines (see Fig. 8.1). The end field measurement was chosen for this demonstration as the end fields vary in the range of ≈1%, while the mid field varies only by ≈0.05%. Thus the measured multipoles are less affected by the accuracy limitations of the measurement devices. Therefore the strengths and weaknesses of a description based on cylindrical elliptical multipoles is demonstrated. The coil probe used for measuring these magnets was long enough (600 mm, Table 5.1) to cover the whole end field so that B_z vanishes at both ends of the probe. Measurement data obtained on the CSLD and S2LD are used to aid the understand-

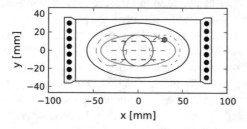

Fig. 8.1 The gap of the magnet (CSLD) and the measurement positions of the coil probe. The eight *black* circles on the *left* and on the *right* indicated the eight turns of the magnet's coil windings. The circles indicate the different positions of the measurements (dashed centre circle, dashed *dotted circle* $x_m = \pm 25$ mm). The *green* ellipse indicates the area of the reconstructed field. The solid dot on the ellipse indicates the angle ψ_p and the cross ψ_c. The *black* ellipse indicates the vacuum chamber's geometry. The three horizontal *dashed lines* indicate the planes, where the field was measured with the hall probe mounted on the mapper

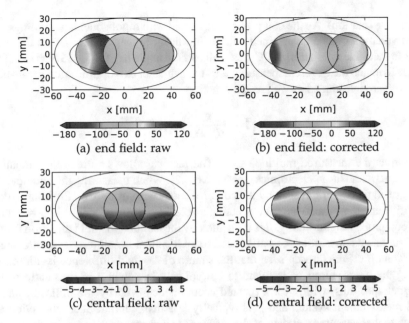

(a) end field: raw

(b) end field: corrected

(c) central field: raw

(d) central field: corrected

Fig. 8.2 Raw and corrected data as measured by the rotating coil probe for the field homogeneity ΔB_y for a field of 0.22 T for the CSLD. The data for the end field are shown in the *top* row while the date for the central field are shown in the *bottom line*. Colour scale in units. 1 unit is equal to 100 ppm

ing, as these magnets were tested using the same method. It is shown in Sect. 8.2.4 that the method is also applicable for the CSLD. In particular the CSLD has been measured with a straight anticryostat, and thus also along a straight line. Hence one can treat the field similar as the field of a straight magnet and consider the curvature effects as local variations of the field.

The S2LD magnet's gap is 125 mm × 68 mm. As the SIS100 magnets are utilising a superconducting coil and a cold iron yoke and, thus, are operated at ≈ 4.5 K, an anticryostat was used to provide room temperature access to the magnet aperture. The anticryostat limits the space accessible within the magnet. The magnetic field was measured at three different lateral positions ($x_m = 0, \pm 30$ mm) with a rotating coil probe of 17 mm radius. The same method was applied to measure the CSLD, here with a spacing of $x_m = 0, \pm 25$ mm (see Fig. 8.1). If the field is then plotted within the circles using the obtained multipole coefficients (see Fig. 8.2a for the end field measurement and Fig. 8.2c for the central field), it is found that the field plots are not continuous in the overlapping area, which were a clear violation of the potential equation or $\Delta B = 0$. The source of this deviation is the limited accuracy of the main field measurement. The field direction was not available for any measurement.

In the following the combination method is described. In a first step all multipoles were rotated to cancel the skew dipole i.e. so that the dipole field of each measurement coincides with the vertical axis of the central measurement (see Sect. 4.1.2 for multipole rotation). As the coil probe is equipped with dipole bucking windings, the field variation can be determined with a significant higher accuracy than the main dipole field (the field strength $|\mathbf{C_1}|$ as well as field angle $arg(\mathbf{C_1})$, see Sect. 5.4, [10, 11]). To match the different measurements to each other two optimisation parameters g and β are introduced as well as the function

$$\mathbf{C_n} = \left(1 + \frac{g}{10\,000}\right) \bar{\mathbf{C}}_n e^{\iota(n+1)\beta} \tag{8.1}$$

to correct the measured multipoles $\bar{\mathbf{C}}_n$. The factor g allows correcting the gain of the signal acquisition electronics (see Sect. 6.3.2). When measuring the end field the integrated field strength seen by the coil probe depends strongly on the longitudinal position and thus g also allows correcting errors due to imperfections in positioning the coil probe longitudinally. β is motivated by the fact that the field direction for the measurements does not necessarily need to be identical for all three measurements. $g^{l,c,r}$ and $\beta^{l,c,r}$ are chosen such that the square of the field difference is minimised along the intersections. g is found to be typically less than 5 in the centre of the magnet and, thus, within the expected accuracy achieved by the utilised rotating coil measurement system and β is typically less than 1 mrad. Such correction are made to the measurement data of the different coil probes (see Fig. 8.2b, d). For the measurement of the magnet end the $g^{l,c,r}$ are found to be in the order of 20. This results from the limited longitudinal coil probe positioning accuracy, as the field is strongly dependent of the longitudinal coordinate at the magnet end. The end field plots show (see Fig. 8.2a, b) that the correction might require large values for $g^{l,r}$. Only after the corrections have been applied the field variation is continuous across all measurements. The effect of these correction steps can be also seen in Figs. 8.3 and 8.4. The central field plots (see Fig. 8.2c, d) show that the corrections of the field are rather small (in the order of 1 unit or less). However, even these small corrections are required. If this discontinuity is not corrected, the harmonics to be obtained later on, are strongly affected by this artifact, as they have to represent this "field discontinuity". After these corrections the fields at the edge of the coil measurements are still discontinuous with steps in the order of up to one unit. Thus, the field data of the two adjacent measurements shall be intermixed by a function λ. Therefore the interpolated field $\mathbf{B_i}(\mathbf{z})$ is given by

$$\mathbf{B_i}(\mathbf{z}) = \lambda \sum_{n=1}^{N_m} \mathbf{C_n^c} \left(\frac{\mathbf{z}}{R_m}\right)^{n-1} + (1-\lambda) \sum_{n=1}^{N_m} \mathbf{C_n^{l,r}} \left(\frac{\mathbf{z} - x_m}{R_m}\right)^{n-1} \tag{8.2}$$

with R_m the measurement radius and N_m the highest order multipole used. N_m was chosen to be 10, as any measured higher order harmonics $\mathbf{C_n^{l,c,r}}$ are in the order of the measurement error. To define λ, the measurement error and thus the weight $w^{l,c,r}$ of the appropriate multipoles at any point $\mathbf{z}$ is to be evaluated (further elaborated in Sect. 9.1). An intuitive estimate is the distance of the point of interest from the centre of rotation of the coil probe, so that

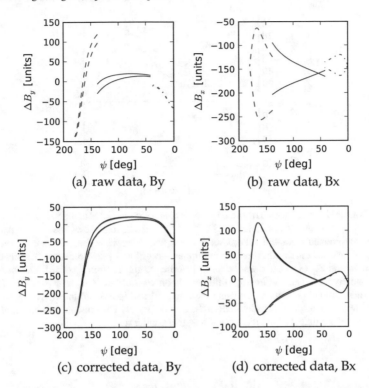

Fig. 8.3 Raw and corrected data as measured by the rotating coil probe for the end field homogeneity ΔB_y and ΔB_x for a nominal field of 0.22 T for the CSLD. All data are plotted along the ellipse versus the angle ψ. For angles from 0 to π the data are represented by *blue lines*, and for angles from $-\pi$ to 0 the data are represented by *red lines*. The *solid lines* represent the data where the multipoles of the central coil probe were used. The *dashed lines* indicate, where the multipoles of the measurement from the *left side* were used, and the *dashed dotted lines* show, where the multipoles of the measurement from the *right side* were used

$$w^l = \frac{R_m}{|\mathbf{z} - x_m|}, \qquad w^c = \frac{R_m}{|\mathbf{z}|}, \qquad \text{and} \qquad w^r = \frac{R_m}{|\mathbf{z} + x_m|}. \qquad (8.3)$$

$\hat{\lambda}^{cl}$ and $\hat{\lambda}^{cr}$ are chosen to

$$\hat{\lambda}^{cl} = \frac{w^c}{(w^c + w^l)} \qquad \hat{\lambda}^{cr} = \frac{w^c}{(w^c + w^r)}. \qquad (8.4)$$

Given that the weights w^l, w^c and w^r are in the order of one and the weight in the area in question is almost linear (see Fig. 8.5), λ is not modelled as given above but chosen to enforce a continuous $\mathbf{B_i(z)}$ and a continuous first derivative of $\mathbf{B_i(z)}$. Thus λ is chosen to be

$$\lambda(p_0) = 0, \quad \lambda(p_1) = 1, \quad \lambda'(p_0) = \lambda'(p_1) = 0, \quad \lambda(p) = 3p^2 - 2p^3,$$

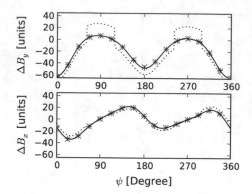

Fig. 8.4 The field as recalculated from the measurements of the end field of the S2LD at an excitation corresponding to a central field level of ≈ 0.22 T. The *dotted line* shows the field as calculated from the measurement data without artifact removal, the *dashed line* the field as calculated form the measurement data after adjusting the multipoles according to the procedure described in the paragraph around Eq. (8.1) and the *solid line* when the weight function λ is used. The *dashed line* coincides nearly with the *solid lines* (a difference is only visible for ΔB_x at $\psi \approx 270°$). The *plus symbols* show the field reconstructed using the elliptical multipoles and the cross symbols when using the circular multipoles calculated from the elliptical ones. The cross and *plus symbol* appear as stars most of the time, as both representations give similar results

with

$$p = \begin{cases} 0 & \psi < p_0 \\ \frac{2\psi - \pi}{2p_0 - \pi} & p_0 \leq \psi \leq \pi, \end{cases} \tag{8.5}$$

with $p_1 = \pi/2$ for the first quadrant. For the other quadrants p can be calculated after reducing the angle to the first quadrant. If one plots the weight functions (Fig. 8.5) one can see that $p_0 = 0.75 \psi_c = \psi_p$ is a better approximation, because then the interpolation functions shape is quite similar to the original weight function (see Fig. 8.1 for ψ_c and ψ_p). Error propagation (see Sect. 9.1) reveals that better results are obtained using $p_0 = \psi_c$. This can be explained by the fact that the distance is proportional to $|z - x_m|$, while for the higher order harmonics the weights are proportional $|(z - x_m)^{n-1}|$. So finally the field can be reconstructed by

$$\mathbf{B(z)} = \begin{cases} \sum_{n=1}^{N_m} \mathbf{C_n^{l,r}} \left(\frac{z \pm a}{R_{Ref}}\right)^{n-1} & \psi \leq \psi_p \\ \mathbf{B_i(z)} & \psi > \psi_p \end{cases} \tag{8.6}$$

$$\psi = \text{Im}\left[\mathbf{cosh}^{-1}(\mathbf{z}/e)\right].$$

The circular multipole coefficients are then calculated as described in Sect. 4.2.2 using this reconstructed field. The results of this approach are demonstrated in Fig. 8.3.

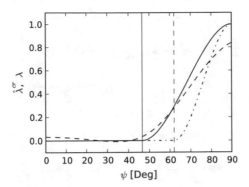

Fig. 8.5 The weight function of the measurement as well as the interpolation functions versus the angle ψ for the first quadrant. The *dashed line* indicates the weight function for the measurement in the centre $2\,\hat{\lambda}^{cr} - 0.5$. The *solid line* indicates the function λ with $p_0 = 0.75\,\psi_c$ and the *dashed dotted line* the function λ with $p_0 = \psi_c$. The scale and offset are used to facilitate the visual comparison to the function λ. The vertical *dashed line* indicates the angle ψ_c and the vertical *solid line* the angle $p_0 = 0.75\,\psi_c$

The elliptical and circular coefficients were now checked interpolating the field on the ellipse and comparing it to the data used for calculating the coefficients. The interpolation error $\delta B_y + \delta i B_x = \mathbf{B_z(z)}$ was defined to

$$\delta \mathbf{B_z(z)} = \frac{10^4}{\mathbf{C_1}}\left[\mathbf{B(z)} - \sum_{n=1}^{N}\mathbf{C_n}\left(\frac{\mathbf{z}}{R_{Ref}}\right)^{n-1}\right] \tag{8.7}$$

for the circular multipoles with N the order of the highest used coefficient and to

$$\delta \mathbf{B_z(z)} = \frac{2 \cdot 10^4}{\mathbf{E_1}}\left[\mathbf{B(z)} - \frac{\mathbf{E_1}}{2} - \sum_{m=2}^{M}\mathbf{E_m}\cosh\left([m-1]\,\mathbf{w}\right)\right] \tag{8.8}$$

$$\mathbf{w} = \cosh^{-1}(\mathbf{z}/e)$$

for the elliptical multipoles with M the order of the highest used coefficient. Figure 8.6 demonstrates that both sets reproduce the field with an acceptable accuracy (in the order of 1 unit). The circular multipoles produce artifacts at the edge of the ellipse if 7 multipoles are used while the elliptical ones describe the fields with nearly no artifacts. Using 30 circular multipoles one can reach a field description as obtained using 8 elliptical coefficients (see also Fig. 8.6). Truncating the sum between the 8th and the 30th multipole will create large artefacts as explained below. Similar results are found on the field representation of the elliptical and circular harmonics for the CSLD.

Even if both sets are able to represent the field with sufficient accuracy the elliptical ones are easier to handle, as the coefficients of the elliptical multipoles get smaller with higher order (see Fig. 8.7). The coefficients of the circular ones on the other

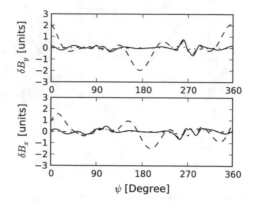

Fig. 8.6 The field interpolation errors for the elliptic representation using the first 8 elliptic coefficients (*solid line*), for the circular representation using the first 7 coefficients (*dashed line*) and for the circular representation using the first 30 coefficients (*dash-dotted line*)

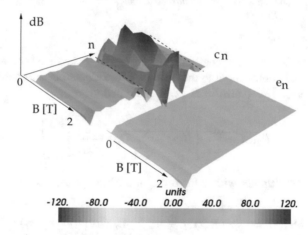

Fig. 8.7 The real part of the elliptical and circular multipole coefficients for different field levels. The biggest coefficients of the $\mathbf{e_n} = 2 \cdot 10^4 \mathbf{E_n}/\mathbf{E_1}$ are found within the leading terms while the circular coefficients c_n show certain bands (indicated by *dashed lines*)

hand produce bands of alternating coefficients. If the sum is now cut within such a band large artefacts might be created. To demonstrate this, the field is reconstructed on the ellipse again for 10 circular multipoles which were calculated using the first 8 elliptical multipoles as described in Sect. 4.2.2. The reconstructed field is given in Fig. 8.8a and compared to the one used to obtain the full set of coefficients (see also Fig. 8.8b). It is found that the artefacts are larger than the field distortion of the magnet itself. Similar effects were also observed for the coefficients calculated for the central field. This clearly shows that the elliptical coefficients are more adapted to describe magnetic fields within an ellipse, but circular multipoles can be still used if the truncation of the series is made with care.

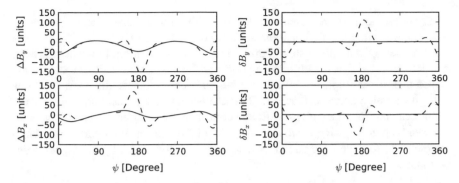

Fig. 8.8 The field as recalculated from the measurement using the elliptical multipoles and circular multipoles for the S2LD next to the interpolation errors. On the *left* the field homogeneity ΔB is plotted versus the angle ψ. The *solid line* shows the field as reconstructed using 8 elliptical multipoles and the *dashed line* as reconstructed using 10 circular multipoles. One can see that the artefacts are nearly as large as the field inhomogeneity itself. On the *right* the field interpolation errors δB are plotted versus the angle ψ for the elliptical (*solid line*) and circular field (*dashed line*) representation for the S2LD. Eight elliptical and ten circular terms were used. The circular multipoles produce artefacts which are larger than the field inhomogeneity itself

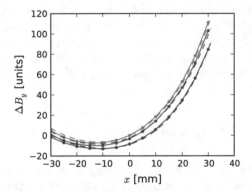

Fig. 8.9 Comparison of the mapper data to the coil probe data for the magnet end. The *dots* indicate the mapper data, the *dashed lines* the reconstructed field using the elliptic multipoles deduced from the coil probe measurements. The *solid lines* are solely used to visually connect the mapper data. In *blue* the data are given for a measurement position y = +10 mm, in *green* for y = 0 and in *red* for y = −10 mm

The data obtained by these measurements were cross checked using a hall probe along a mapper and scanning the field along the lines presented in Fig. 8.1. The comparison of these two measurements is presented for the end field of the CSLD dipole in Fig. 8.9.

One can see that the two measurements match well. This documents that the method described here is sound and reliable and can be used for measuring accelerator magnets.

The difference in the representation quality of the elliptical and circular multipoles is an empirical finding. The analytic correspondence outlined in Sect. 4.2.2 suggests that both sets of multipoles should give an equal representation. Practice, however, has demonstrated that the elliptical multipoles are easier to use simply due to the reason that, if the elliptical multipoles are affected by artefacts, the basis functions will give a value between 0 and 1. On the other hand, the basis functions of the cylindrical multipoles are giving different contributions to different parts of the field. The central part ($x < R_{Ref}$) will be dominated by the lower order harmonics while the part near to the ellipse end ($x > R_{Ref}$) will be dominated by the higher order harmonics.

8.1.2 Measurement Results

Next to the coil probe measurements made with the mole, the field of the magnet S2LD was measured with a hall probe mounted on a mapper. This hall probe showed an absolute accuracy of better than 10^{-4} at 1 T. The transfer function as measured by the probe in the centre as well as measured by the mole is given in Fig. 8.10b [12]. One can hardly notice any difference between the two, thus even in absolute values the two measurement systems agree well. The field harmonics of the S2LD are presented in Fig. 8.10. One can see that the harmonics found in the three measurements agree well (see Fig. 8.10) and differ only at field values where the magnet yoke gets saturated.

8.2 Measuring Toroidal Multipoles

The aperture of bending magnets (i.e. a dipole) in an accelerator can be reduced if the magnet is curved and thus it follows the sagitta of the beam. These magnets are typically measured with search coils, i.e. coil probes which follow the magnets curvature. These were used to measure e.g. SIS18 magnets [13] the magnets of HIT [14] or CNAO [15]. Multipoles are not straight forward to derive from these measurements. Further the coil probe must be aligned with the mid-plane. Here rotating coil probes are applied for measuring curved magnets.

Normally rotating coils are treated in a straight cylinder, thus the longitudinal axis of the coordinate system coincides with the axis rotation (see Chap. 5), which is also used for slightly bent magnets (e.g. LHC [16]). The toroidal circular multipoles (see Sect. 4.3) allow deriving the limits of a rotating coil measurement. The results given here are based on [9, 17, 18]. The integrations are made similarly as for a straight magnet, (see Chap. 5, (5.3))

$$\Phi = \int_{r_1}^{r_2} \int_0^L \vec{B}^t(x, y) \, d\vec{\sigma} \tag{8.9}$$

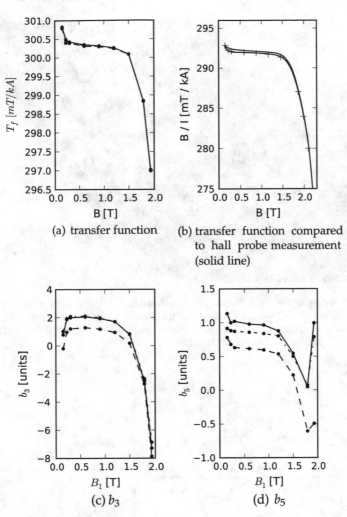

Fig. 8.10 The measured transfer function and harmonics versus the main field as measured for the S2LD. Coil measurements are indicated by *dots* or *crosses*. The *solid line* represents the data for a longitudinal position $z = 0$, the *dashed line* for a position $z = 600$ mm and the *dashed dotted line* for a position of $z = -600$ mm. The coil probe measurements of the different positions match for the transfer function and b_3 so closely, so that they are hardly distinguishable. The data obtained with the hall probe are represented by lines in the *top right* figure and the coil probe measurements by crosses '+'. Figure 8.10 a, b show both the transfer function of the magnet. The transfer function T_f is given by $T_f = B/I$ with B the magnetic induction and I the current

but now the dependence of $\vec{B}^t(x, y, z)$ versus z has to be taken into account [17]. Here one assumes that the field is constant versus the toroidal angle, with the multipoles as given in (4.85) (see Fig. 8.11).

Fig. 8.11 The rotating coil within a torus. The torus has been cut open so that the coil is visible. The *centre line* of the torus is indicated together with the rotation axis of the coil. The coil is indicated as a single turn. The rotation axis is a tangent to the torus. The centre of the rotating coil is also the point where the tangent touches the centre circle

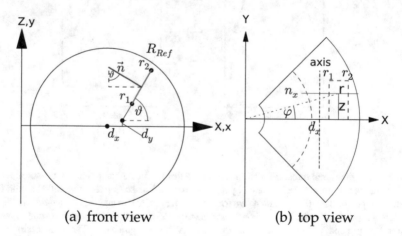

Fig. 8.12 Illustration of the coil in the sector and the integration paths. The *left* figure shows the view from the *front* and the *right* the view from the *top* for the plane $\varphi = 0$. r_1 and r_2 are the inner and outer radius of the coil probe, d the offset of the coil probe rotation axis from the ideal axis, z the longitudinal offset in the local coordinate frame x, y, z

Local toroidal circular coordinates are used to describe the magnetic field (see Sect. 3.3.2 for the coordinates and Sect. 4.3 for the multipoles). The analysis given here is targeted for measuring these local toroidal multipoles, thus terms of $\mathcal{O}(\varepsilon^2)$ or above are neglected, as have been neglected in the differential Eq. (4.72). Different coordinate systems are required to analyse this problem (see Fig. 8.12). The whole calculation shall be made in the local Cartesian coordinate system x, y, z or in an equivalent frame ρ, ϑ, z. Further displacement of the coil probe's centre of rotation shall be taken into account in x-direction by d_x and in y-direction by d_y. The calculation starts at the middle of the coil probe, which shall coincide with the plane, where the toroidal angle $\varphi = 0$. The relation to the global coordinates is then given by

$$X = (R_C + x) \cos \phi = R_C + d_x + r \cos \vartheta, \tag{8.10}$$

$$Y = (R_C + x) \sin \phi = -z, \tag{8.11}$$

$$Z = y = d_y + r \sin \vartheta. \tag{8.12}$$

These are required to derive the local area normal on the coil next to the offset of the torus centre circle from the rotation axis. Equations (8.10) and (8.11) give the relation for φ by

$$\tan \varphi = \frac{Y}{X} = \frac{-z}{R_C + d_x + r \cos \vartheta}, \tag{8.13}$$

which in turn yields

$$\cos \varphi = \frac{1}{\sqrt{1 + \tan^2 \varphi}} = \frac{R_C + d_x + r \cos \vartheta}{\sqrt{(R_C + d_x + r \cos \vartheta)^2 + z^2}}. \tag{8.14}$$

Substituting $R_C = R_{Ref} / \varepsilon$ one can express $\cos \varphi$ by

$$\cos \varphi \approx 1 - \varepsilon^2 \frac{z^2}{2 R_{Ref}^2} \approx 1 + \mathcal{O}(\varepsilon^2). \tag{8.15}$$

Equations (8.10) and (8.14) allow expressing the x coordinate of the rotating coil probe frame in the local Cartesian frame moving along the larger torus circle

$$x = \frac{R_C + d_x + r \cos \vartheta}{\cos \varphi} - R_C = \sqrt{(R_C + d_x + r \cos \vartheta)^2 + z^2} - R_C$$

$$= R_C \left[1 + \frac{d_x + r \cos \vartheta}{R_C} + \frac{z^2}{2 R_C^2} + \cdots - 1 \right] \tag{8.16}$$

in dependence of z. It can be approximated by

$$x = \chi_x (r, \vartheta, z) := d_x + r \cos \vartheta + \varepsilon \frac{z^2}{2 R_{Ref}} + \mathcal{O}(\varepsilon^2), \tag{8.17}$$

which approximates the larger torus circle with a parabola. From Eq. (8.12) one can derive y to

$$y = \chi_y (r, \vartheta, z) := d_y + r \sin \vartheta. \tag{8.18}$$

8.2.1 The Magnetic Flux

For calculating the flux Φ (8.9) the surface normal has to be calculated in the polar frame of the coil probe r, ϑ. In the rotating coil probe frame the surface normal $\vec{n}'$ can be expressed by the vector in the local Cartesian frame using (8.17) and (8.18) as

$$\vec{n}' = (-\sin\vartheta', \cos\vartheta', 0), \qquad \vec{n} = (-\sin\vartheta\cos\varphi, \cos\vartheta, \sin\vartheta\sin\varphi) \quad (8.19)$$

(see also Fig. 8.12a). It is assumed that the magnetic field is invariant versus φ. So the field components B'_x and B'_y, B'_z in the coil probe frame are expressed by (see Fig. 8.12b)

$$\vec{B}' = (B'_x, B'_y, B'_z) = (B_x\cos\varphi, B_y, B_x\sin\varphi). \qquad (8.20)$$

in the local Cartesian frame. B'_z can be ignored as the rotating coil probe is insensitive to this component. Then the field is given by

$$\vec{B}_n(x, y) = (\vec{n}' \cdot \vec{B}') = (\vec{n} \cdot \vec{B}) =$$
$$= -B_x\big(\chi_x(r, \vartheta, z), \chi_y(r, \vartheta, z)\big)\sin\vartheta\cos\varphi + B_y\big(\chi_x(r, \vartheta, z), \chi_y(r, \vartheta, z)\big)\cos\vartheta. \qquad (8.21)$$

Here $\cos\varphi \approx 1$ is used again. Now the local toroidal circular multipoles (4.85) are inserted for $\vec{B}$ in (8.21) and x and y are described by (8.17) and (8.18) which yields

$$\vec{B}_n = \sum_{m=1}^{M}\left[R_m\left(\vec{T}_m^{\,n}(\chi_x, \chi_y)\cdot\vec{n}\right) + S_m\left(\vec{T}_m^{\,s}(\chi_x, \chi_y)\cdot\vec{n}\right)\right]. \qquad (8.22)$$

The field is integrated over r and z, which gives the flux

$$\Phi(\vartheta) = \int_0^L\int_{r_1}^{r_2}\vec{B}_n(r, \vartheta, z)\,\mathrm{d}r\,\mathrm{d}z \qquad (8.23)$$

seen by the coil probe at the angle ϑ. This integral is evaluated for the different multipoles of order m up to order $M = 20$ using Mathematica™ (see Sect. C.1). 20 multipoles are typically more than sufficient for representing the field of accelerator magnets.

For the following calculations the index variable m is now replaced by μ and the index variable n by ν so that these are not confused with the upper indices. To interpret the result of the integral (8.23) one decomposes the flux generated by each local toroidal multipole into the cylindrical circular multipoles, as the flux seen by the coil probe must be independent of the description used for representing the magnetic field. The flux seen by a rotating coil probe (see 5.11) is given by

$$\Phi = \mathrm{Re}\left(\sum_{\nu=1}^{N} \mathbf{K}_\nu \left[B_\nu + \mathrm{i}A_\circ\right] e^{\iota \nu \vartheta}\right) = \sum_{\nu=1}^{N} K_\nu \left[B_\nu \cos \nu\vartheta - A_\nu \sin \nu\vartheta\right], \quad (8.24)$$

if a cylindrical circular description is used. Here a radial coil probe is treated and thus K_ν is real (see Sect. 5.2). The comparison is made by the following steps. The result of this integral is evaluated for each μ using

$$R_\mu N_w \int_0^{L}\int_{r_1}^{r_2} (\vec{T}_\mu^{\mathrm{n}}(\chi_x, \chi_y)\cdot \vec{n})\, \mathrm{d}r\, \mathrm{d}z = \sum_{\nu=1}^{M+1}\left[G'^{nn}_{\mu,\nu}\, K_\nu\, \cos(\nu\vartheta) + G'^{ns}_{\mu,\nu}\, K_\nu\, \sin(\nu\vartheta)\right],$$

$$S_\mu N_w \int_0^{L}\int_{r_1}^{r_2} (\vec{T}_\mu^{\mathrm{s}}(\chi_x, \chi_y)\cdot \vec{n})\, \mathrm{d}r\, \mathrm{d}z = \sum_{\nu=1}^{M+1}\left[G'^{ss}_{\mu,\nu}\, K_\nu\, \sin(\nu\vartheta) + G'^{sn}_{\mu,\nu}\, K_\nu\, \cos(\nu\vartheta)\right],$$

$$\mu = 1, 2, ..., M. \quad (8.25)$$

The integral gives the area which corresponds to a coil probe with one turn, N_w represents the appropriate number of turns, which is part of the standard definition of the coil probe sensitivity (see (5.13) or (8.24)). The integrals on the left give the flux which is generated by each individual toroidal multipole. The right side shows the flux as seen by a rotating coil probe for cylindrical circular multipoles (see (8.24)) substituting either B_n with G'^{nn} and A_n with G'^{ns} or B_n with G'^{sn} and A_n with G'^{ss}. Thus G'^{nn}_μ gives the normal coil probe signal B_ν which are induced by a normal toroidal multipole R_μ, G'^{ns} the skew coil probe signals A_ν which are induced by a normal toroidal multipole R_μ and so forth. The vector G' is constructed for each toroidal multipole R_μ and S_μ. Then these vectors are combined to matrices which can be split into

$$\begin{aligned} G^{nn}_{\mu,\nu} &= \delta_{\mu\nu} + H^{nn}_{\mu,\nu}, & G^{ns}_{\mu,\nu} &= & H^{ns}_{\mu,\nu}, \\ G^{sn}_{\mu,\nu} &= & H^{sn}_{\mu,\nu}, & G^{ss}_{\mu,\nu} &= -\delta_{\mu,\nu} + & H^{ss}_{\mu,\nu}, \\ & & \mu, \nu = 1, 2, ..., M, & & \end{aligned} \quad (8.26)$$

as the absolute value of all off diagonal elements is much smaller than 1. Each of this matrices G and H is of size $M \times (M+1)$; for further treatment it was limited to a $M \times M$ matrix. The column $M + 1$ contained only one element and will be discussed end of Sect. 8.2.2. These matrices are now combined to one single $2M \times 2M$ matrix with

$$G := G^0 + H = G^0 + \begin{pmatrix} H^{nn} & H^{ns} \\ H^{sn} & H^{ss} \end{pmatrix} \quad (8.27)$$

G^0 is a $2M \times 2M$ digonal matrix. The diagonal comprises M 1's and then M -1's, hence

$$G^0 = \mathrm{dia}(1, 1, 1, \ldots, -1, -1, -1, \ldots) = (G^0)^{-1}. \quad (8.28)$$

It shows that a normal toroidal multipole generates a signal quite similar to a normal cylindrical multipole and a skew toroidal multipole generates mainly a cylindrical multipole. If the absolute value of the elements of G is much smaller than 1, its inverse can be approximated by (see Appendix D).

$$G^{-1} \approx G^0 - G^0 \cdot H \cdot G^0. \tag{8.29}$$

8.2.2 Conversion Matrices

The sections given above showed that the toroidal multipoles can be deduced from rotating coil probe measurements assuming that the field is invariant versus φ. The calculations above also considered the effect of a misplaced coil. The cylindrical circular multipoles are mapped to the toroidal circular multipoles by

$$\begin{pmatrix} \vec{R}_\mu \\ \vec{S}_\mu \end{pmatrix} = \begin{pmatrix} G^{nn}_{\nu,\mu} & G^{ns}_{\nu,\mu} \\ \hline G^{sn}_{\nu,\mu} & G^{ss}_{\nu,\mu} \end{pmatrix} \begin{pmatrix} \vec{B}_\nu \\ \vec{A}_\nu \end{pmatrix}. \tag{8.30}$$

Each of the submatrices G^{nn}, G^{ns}, G^{sn} and G^{ss} is set up by

$$G = I + \mathcal{L}^{dr} + \varepsilon \left(\mathcal{L}^L + U + \mathcal{L}^{sk} + \mathcal{L}^{R2} + \mathcal{L}^{R2_0} \right); \tag{8.31}$$

but not each of them contains all terms of the equation above. The full solution is given in (8.41). Only elements of $\mathcal{L}^L$ depend on the coil length L while only elements of $\mathcal{L}^{sk}$ depend on the coil sensitivity parameters (see 5.12). The elements of the remaining matrices depend only on d_x and d_y. All these matrices can be derived from complex matrices, but the result itself is not an analytic complex function. In the following part the coefficients of the different submatrices will be given.

The matrix U is the sole one which only consists of constant terms and is given by

$$U = \frac{\nu + 1}{4\nu} \delta_{\nu+1,\mu}. \tag{8.32}$$

Some of the matrices below are given as triangular lower matrices. Therefore one defines

$$\mathcal{L}_{\nu,\mu} = \begin{cases} 1 & \nu \geq \mu \\ 0 & \nu < \mu. \end{cases} \tag{8.33}$$

Similar to measuring with rotating coil probes, an offset of the coil probe causes that one multipole creates spurious other multipoles. These are similar for the different submatrices of matrix H and thus summarised here. The matrix $\mathcal{L}^{dr}$ is the only one, which does not depend on the torus curvature ratio ε. Its non zero elements are given by

$$\mathcal{L}^{dr}_{\nu,\mu} = \begin{pmatrix} \nu - 1 \\ \mu - 1 \end{pmatrix} * \left(\frac{\mathbf{d_z}}{R_{Ref}} \right)^{\nu - \mu} * \mathcal{L}_{\nu,\mu} - I \,. \tag{8.34}$$

with $\mathbf{d_z} = d_x + \mathrm{i} d_y$ and I the identity matrix. The "*" denotes that these multiplication is to be made element wise. This term is due to the frame translation in d_x and d_y, which is exactly the same as found if a rotating coil probe is displaced by $d_x + \mathrm{i} d_y$ within a cylindrical circular coordinated system. This effect is called the "feed-down" effect. The identity matrix ($I = \delta_{\nu,\mu}$) is subtracted as the diagonal has to be singled out for later treatment.

A utility matrix $\mathcal{L}^{ddr}$ needs to be defined, which in turn requires to define $\mathcal{L}^{d}$, which is given by

$$\mathcal{L}^{d}_{\nu,\mu} = \begin{cases} 1 & \nu > \mu + 1 \\ 0 & \nu \leq \mu + 1 \,. \end{cases} \tag{8.35}$$

Then $\mathcal{L}^{ddr}$ is given by

$$\mathcal{L}^{ddr}_{\nu,\mu} = (\nu - \mu) * \begin{pmatrix} \nu - 1 \\ \mu - 1 \end{pmatrix} * \left(\frac{\mathbf{d_z}}{R_{Ref}} \right)^{\nu - \mu - 1} * \mathcal{L}^{d}_{\nu,\mu} + \mu \delta_{\nu,\mu+1}$$

$$= R_{Ref} \frac{\mathrm{d}}{\mathbf{d_z}} \mathcal{L}^{dr} \tag{8.36}$$

with $\mathbf{z} = x + \mathrm{i} y$ and is similar to (8.34) except for the binomial factor and that the power is reduced by 1. Please note, that the band below the diagonal (i.e. $\mathcal{L}^{ddr}_{\nu,\mu} * \delta_{\nu,\mu+1}$) does not depend on $\mathbf{d_z}$. The dependence on L is given by

$$\mathcal{L}^{L}_{\nu,\mu} = \frac{L^2}{3 R^2_{Ref}} \mathcal{L}^{ddr} \,. \tag{8.37}$$

The dependence on the coil sensitivity factors K_μ is given by

$$\mathcal{L}^{sk}_{\nu,\mu} = \frac{1}{4 (\mu + 1)} * \frac{K_{\mu+2}}{K_\mu} * \mathcal{L}^{ddr} \,. \tag{8.38}$$

The coil sensitivity K is real here as only radial coil probes are considered (see also 5.13). The description of $\mathcal{L}^{R2}_{\nu,\mu}$ requires considerably more terms than any of the other submatrices. It only depends on d_x and d_y. It is given by

$$\mathcal{L}^{R2}_{\nu,\mu} = \frac{1}{4\nu} \left(\frac{\nu, \mu}{\nu - \mu + 1} * \mathcal{L}_{\nu,\mu} + \delta_{\mu,1} \right) *$$

$$* \left[\frac{d_y}{R_{Ref}} - \left(\frac{2 - \mu + 2\nu}{\mu} * \mathcal{L}_{\nu,\mu} - \nu \delta_{\mu,1} \right) \mathrm{i} \frac{d_x}{R_{Ref}} \right] * \left(\mathcal{L}^{dr} + I \right) \,. \tag{8.39}$$

The last matrix $\mathcal{L}^{R2_0}$ is given by

$$\mathcal{L}^{R2_0}{}_{\nu,\mu} = \frac{1}{2\nu} \left(\frac{\mathbf{d_z}}{R_{Ref}} \right)^{\nu} \delta_{\mu,1}. \tag{8.40}$$

Each of the submatrices G^{nn}, G^{ns}, G^{sn} and G^{ss} is set up by

$$
\begin{aligned}
G^{nn} &= I + \mathrm{Re}\left[\mathcal{L}^{dr}\right] + \varepsilon\left(-U + \mathrm{Re}\left[\mathcal{L}^{L}\right] - \mathrm{Re}\left[\mathcal{L}^{sk}\right] + \mathrm{Im}\left[\mathcal{L}^{R2}\right] + \mathrm{Re}\left[\mathcal{L}^{R2_0}\right]\right), \\
G^{ns} &= -\mathrm{Im}\left[\mathcal{L}^{dr}\right] + \varepsilon\left(-\mathrm{Im}\left[\mathcal{L}^{L}\right] + \mathrm{Im}\left[\mathcal{L}^{sk}\right] + \mathrm{Re}\left[\mathcal{L}^{R2}\right]\right), \\
G^{sn} &= -\mathrm{Im}\left[\mathcal{L}^{dr}\right] + \varepsilon\left(-\mathrm{Im}\left[\mathcal{L}^{L}\right] + \mathrm{Im}\left[\mathcal{L}^{sk}\right] + \mathrm{Re}\left[\mathcal{L}^{R2}\right] - \mathrm{Im}\left[\mathcal{L}^{R2_0}\right]\right), \\
G^{ss} &= -I - \mathrm{Re}\left[\mathcal{L}^{dr}\right] + \varepsilon\left(+U - \mathrm{Re}\left[\mathcal{L}^{L}\right] + \mathrm{Re}\left[\mathcal{L}^{sk}\right] - \mathrm{Im}\left[\mathcal{L}^{R2}\right]\right).
\end{aligned}
\tag{8.41}
$$

This summary already shows that the main dipole is affected by all measured harmonics. On the other hand ε is rather small for the machines considered here. The different matrices H are obtained omitting the identity matrices. Comparing the operators Re and Im on $\mathcal{L}^{L}$ and $\mathcal{L}^{sk}$ to the ones operating on $\mathcal{L}^{dr}$ in (8.41) one can assume that $\mathcal{L}^{L}$ and $-\mathcal{L}^{sk}$ are analytic. These matrices can also be used to construct the integrals (8.25); then one more term of the matrix U has to be taken into account than for the other matrices.

8.2.3 Choosing a Coil Probe Length

Evaluating all the different terms one can see that only $\mathcal{L}^{L}$ is of significant size for accelerator magnets with characteristic values as given in Table 8.1. A criterion can be given for defining an adequate coil probe length by demanding that the feed down effect as found for cylindrical circular multipoles and as found for toroidal circular multipoles should be of equivalent size.

To derive this criterion one defines

$$\mathcal{L}^{dL}_{\nu,\mu} = \begin{cases} 1 & \nu > \mu \\ 0 & \nu \leq \mu + 1. \end{cases} \tag{8.42}$$

Then the criterion is given by

Table 8.1 Geometries of different accelerators

	R_C (m)	R_{Ref} (mm)	ε (units)	L (mm)	$|\mathbf{d_z}|^{max}$ (mm)
LHC	2804	17	0.04	600	1
SIS100	52.625	40	7.62	600	1
SIS300	52.625	35	6.67	600	1
NICA	15	40	26.67	600	1

$$\mathcal{L}_{\nu,\mu}^{dr} = \underbrace{\frac{3R_{Ref}^2}{\varepsilon L^2} \frac{\mathbf{d_z}}{R_{Ref}}}_{L_s} \left(\frac{1}{\nu - \mu} * \mathcal{L}_{\nu,\mu}^{dL} \right) * \mathcal{L}_{\nu,\mu}^L . \tag{8.43}$$

The feed down effect $\mathcal{L}^{dr}$ creates spurious harmonics; for $\mathcal{L}^{dr}$ the spurious harmonic due to the next higher harmonic will be proportional to $\mathbf{dz}/R_{Ref}$. On the contrary for $\mathcal{L}^L$ the spurious harmonic generated by the next higher harmonic will only depend on the coil length L and the large circle radius of the torus $R_C = \varepsilon R_{Ref}$. Only the spurious harmonic due to the next but one harmonic will depend on $\mathbf{dz}/R_{Ref}$. This allows to draw two different conclusions:

- one demands that the spurious harmonics due to $\mathcal{L}^L$ shall not be bigger than a maximum acceptable displacement $|\mathbf{d_z}|^{\max}$, thus $L_s := 1$. Then the maximum coil length L^{max} can be deduced to

$$L^{max} = \sqrt{\frac{3\,|\mathbf{d_z}|^{\max}\,R_{Ref}}{\varepsilon}} = \sqrt{3\,|\mathbf{d_z}|^{\max}\,R_C}. \tag{8.44}$$

- Or one accepts the spurious harmonics created by $\mathcal{L}^L$ independent of $\mathbf{dz}$ and relates the terms to each other only for the terms dependent on $\mathbf{dz}$. Then one deduces L_2^{max} dividing L_s with $\mathbf{dz}/R_{Ref}$ which yields

$$L_2^{max} = \sqrt{3\,R_C\,R_{Ref}}, \tag{8.45}$$

which will give a much larger acceptable coil length as (8.44). The definition of L_2^{max} can give so large values that one should first deliberate whether the approximations of $\mathcal{O}(\varepsilon)$ are still valid in all the calculations given above. Further one should investigate, if the coil probe will cover the area of interest to an acceptable extent as the offset of the centre circle from the central toroidal circle could get really large (see (8.17) for its dependence on z).

The values of L_2 obtained for the geometries of the machines listed in Table 8.1 are so large that this definition was abandoned.

So the definition of (8.44) is further investigated. The factor $1/(\nu - \mu)$ in (8.43) shows that the spurious harmonics due to $\mathcal{L}^L$, (8.37), will be larger than those due to $\mathcal{L}^{dr}$, (8.34), when one considers the feed-down effect of harmonics which are off by more than 1. But for these the offset $\mathbf{d_z}$ scales with $(\mathbf{d_z}/R_{Ref})^{\nu-\mu}$ and thus will decrease much faster than the term $1/(\nu - \mu)$ will grow. To understand the effect (8.44) $|\mathbf{d_z}|^{\max}$ has to be investigated. This maximum error is understood here as an error which can be dealt with: either the spurious harmonics are small enough so that these can be ignored or it can be corrected for. If a larger $|\mathbf{d_z}|^{\max}$ is acceptable in the measurement also a longer coil probe length can be handled; in particular the largest contribution will be due to the factors independent of $\mathbf{dz}$. These factors depend only on the coil probe length L (as some of the coefficients in $\mathcal{L}^L$ are independent of $\mathbf{dz}$) and thus depend only on the coil probe's geometry. These factors are large and can be ignored in a dipole as the part of the basis functions of the local toroidal multipoles depending on ε is 0 for the first term ("dipole") but not for the second term ("quadrupole"). Thus a large influence of $\mathcal{L}^L$ is expected if a combined function magnet (i.e. a dipole and a quadrupole in one magnet) is to be measured with a rotating coil probe.

The attention of readers familiar to coil probes and evaluating their measurements shall be drawn to the influence of the sensitivity factors (see (8.38)). The first term affecting the "dipole" is the "sextupole" term. Here the ratio can be very small, if compensating systems or "bucking" systems are used (see e.g. [10]). Any further treatment will require to invert the matrices (8.41); in this case the sign of the term will swap and its magnitude change should be rather small, given that the identity matrix is involved (see (8.27) and (8.29)).

8.2.4 Magnitude of the Terms

The formulae given above were evaluated for the geometries of the following different machines: the Large Hadron Collider (LHC) at CERN[19], SIS100 [20, 21] and SIS300 at GSI, and NICA [22, 23] at Dubna (see Table 8.1). The parameters given in Table 8.1 were used to calculate the coefficients of the matrices. Accelerators require a field description with an accuracy of 1 unit and roughly 0.1 unit for the field homogeneity (1 unit equals 100 ppm). Therefore any contribution less than 1 ppm can be ignored.

Due to the circumference of the LHC ε is very small and thus the correction of all matrices are very small (less than 1 ppm) except for the matrix $\mathcal{L}^L$, where the values close to the diagonal get to a size of 2000 ppm for $|\mathbf{d_z}| = R_{Ref}$. So even for an insane value of $\mathbf{d_z}$ the artefacts can be handled. This value may seem to exceed the target value for the field description; but the higher order multipoles are in the order of 100 ppm; thus the effective artefact will be safely below the target value of 10 ppm.

For machines with an aspect ratio as found for SIS100 or SIS300 the matrix U is in the order of 100 ppm. It can be neglected except for the main multipole. The values of the matrix $\mathcal{L}^L$ get of similar size as the values for $\mathcal{L}^{dr}$. The magnitude of this value is defined by the magnitude of the offset $|\mathbf{d_z}|$. Also when measuring straight magnets special methods are applied to obtain the offset $\mathbf{d_z}$ from the measured dataset [10]. Therefore one can assume that the artefacts can be minimised by similar suitable procedures.

The parameters for the different machines are given in Table 8.1. An appropriate coil length was deduced imposing that the influence of the offset of the coil from the tangent of the larger torus circle shall be of the same order as for a coil probe measuring a straight magnet (see (8.44)). The matrices (8.41) list all toroidal multipoles required to describe the field for one measured multipole B_ν or A_ν. Now the matrices are evaluated to deduce for each multipole the strength of the corresponding toroidal multipoles R_μ or S_μ. For the geometry of the machines considered here, as listed in Table 8.1, only the terms U and $\mathcal{L}^L$ give a significant contribution; thus only the expression $U + \mathcal{L}^L$ is evaluated below. It is given by

$$C = U + \mathcal{L}^L \tag{8.46}$$

$$= \begin{pmatrix}
\cdot & \frac{1}{2} & & & & & \\
\frac{L^2}{3R_{Ref}^2} & \cdot & \frac{3}{8} & & & & \\
\frac{2L^2 d_z}{3R_{Ref}^3} & \frac{2L^2}{3R_{Ref}^2} & \cdot & \frac{1}{3} & & & \\
\frac{L^2 d_z^2}{R_{Ref}^4} & \frac{2L^2 d_z}{R_{Ref}^3} & \frac{L^2}{R_{Ref}^2} & \cdot & \frac{5}{16} & & \\
\frac{4L^2 d_z^3}{3R_{Ref}^5} & \frac{4L^2 d_z^2}{R_{Ref}^4} & \frac{4L^2 d_z}{R_{Ref}^3} & \frac{4L^2}{3R_{Ref}^2} & \cdot & \frac{3}{10} & \\
\frac{5L^2 d_z^4}{3R_{Ref}^6} & \frac{20L^2 d_z^3}{3R_{Ref}^5} & \frac{10L^2 d_z^2}{R_{Ref}^4} & \frac{20L^2 d_z}{3R_{Ref}^3} & \frac{5L^2}{3R_{Ref}^2} & \cdot &
\end{pmatrix}.$$

The dots indicate the diagonal, where all elements are zero. The parameters given in Table 8.1 were inserted. The matrix C is inverted which gives

$$C^{-1} = I + \frac{1}{10000} \begin{pmatrix}
\cdot & 4 & & & & & \\
143 & \cdot & 3 & & & & \\
3\,285 & \cdot & 3 & & & & \\
& 9\,428 & \cdot & 2 & & & \\
-1 & 18\,570 & \cdot & & & & \\
-2 & 31\,713 & \cdot & & & & \\
-3 & 46 & & & & & \\
-6 & & & & & &
\end{pmatrix}, \tag{8.47}$$

with all elements rounded to 1 unit. Elements smaller than one unit were left out. The dots indicate again the diagonal. The higher order harmonics, measured with the coil probe are in the order of some units. The basis terms of the toroidal circular multipoles (see (4.128) are scaled with $\varepsilon/4$ and the magnitude of term T_1 and T_2 is still less than 2. For accelerator magnets one can safely assume that all higher order harmonics are well below 10 units. So one can conclude that the effect of this matrix can be neglected for all measured harmonics except the main one if an field description accuracy of not better than $\frac{\varepsilon}{4}\frac{10}{10\,000}$ is required. The toroidal circular term (see Table 4.4 and (4.128)) for $m = 1$ gives also a quadrupole and a component caused by term T_2; thus the perturbation term is then exactly zero for the normal part. The skew part is $4x/R_{Ref}$ (see Table 4.4), but this can be neglected as the skew component is small (<10 units) and still has to be multiplied with ε. Therefore only a quadrupole of $\approx140 \approx20\,\varepsilon$ units and a sextupole of ≈3 units has to be added to the set of cylindrical circular multipoles. Then the cylindrical circular multipole description can be used.

A measurement procedure for obtaining elliptical circular multipoles was given in Sect. 8.1, with the measurements performed at different circles: one was measured in the centre of the magnet and the others shifted by $\pm x_m$. For the measurement of the S2LD $x_m \pm 30$ mm and for the measurement of the CSLD $x_m \pm 25$ mm was used. So one can define

$$R_C^{\pm} = R_C \pm x_m \quad \text{and} \quad \varepsilon^{\pm} = \frac{R_{Ref}}{R_C \pm x_m}. \tag{8.48}$$

Using $x_m \pm 30$ mm the different ε are then given (in units) by

$$\varepsilon^{\pm} = 7.601 \pm 0.004 \tag{8.49}$$

using the values for SIS100 given in Table 8.1. The change of ε is at the 7th digit and is thus significantly smaller than the measurement accuracy obtainable with the systems given here. This result shows that the cylindrical elliptical multipoles can be used to treat the measurements of the curved dipole magnets of SIS100.

The calculations for SIS300 yield a matrix with numerical values of

$$C_{SIS300}^{-1} = I + \frac{1}{10000}
\begin{pmatrix}
 & 3 & & & & & & \\
125 & . & 2 & & & & & \\
 & 3\,249 & . & 2 & & & & \\
 & & 9\,374 & . & 2 & & & \\
 & & & 19\,499 & . & & & \\
 & & & & -1 & 31\,624 & & \\
 & & & & & -2 & 47 & \\
 & & & & & & -4 &
\end{pmatrix} \tag{8.50}$$

thus the effect of curvature can be neglected for a coil probe length of 600 mm, if the quadrupole is recalculated. The SIS300 magnets have a round aperture; thus different coil positions do not need to be evaluated.

For the NICA machine [22, 23] the inverse of matrix C_{NICA} is given by

$$C_{NICA}^{-1} = I + \frac{1}{10000}
\begin{pmatrix}
-1 & 13 & & & & & & \\
500 & -2 & 10 & & & & & \\
-25 & 1000 & -2 & 9 & & & & \\
1 & -75 & 1500 & -3 & 8 & & & \\
 & 4 & -150 & 2001 & -4 & & & \\
 & & 9 & -250 & 2501 & & & \\
 & & & 19 & -375 & & & \\
 & & & 1 & 33 & & &
\end{pmatrix}, \tag{8.51}$$

which shows that the effects roughly increase with l/R_C'. Using the same x_m, but $R_C' = 15$ m, one gets

$$\varepsilon_{NICA}^{\pm} \approx 26.67 \pm 0.053, \tag{8.52}$$

with ε_{NICA} in units. The values of the matrix C_{NICA}^{-1} are roughly three times higher than for C^{-1} (SIS100). Similarly ε_{NICA} is an order of magnitude larger than ε for SIS100. This influence will have to be evaluated and compared to the required field quality descriptions to see if the evaluation using circular multipoles is still precise enough.

8.2.5 Measurement Results on the SIS100 Curved Dipole Magnet

Following the above reasoning (see Sect. 8.2.4) the magnetic field of a curved SIS100 dipole can be described with sufficient accuracy with cylindric elliptical and circular multipoles, using appropriate recalculations. A curved SIS100 dipole model magnet was built (see Sect. 1.3) and the field was measured using the mole (see Sect. 6.3.2, [5, 20]).

The field is described with plane elliptical multipoles, which are then transformed to circular ones (see Sect. 4.2.2). The reference radius R_{Ref} is 40 mm. The first allowed harmonics are given in Fig. 8.13. These measured multipole coefficients are the basis of the evaluation of the field quality of the CSLD.

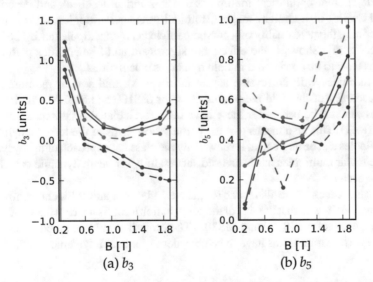

(a) b_3 (b) b_5

Fig. 8.13 The first two allowed harmonics measured on the CSLD. The strength of the multipole (in units) is given versus the main field (in Tesla). The different longitudinal positions are indicated by *colour* and *line*. The centre is indicated in *red* ($z = 0$), *green* for $z = \pm 300$ mm, (*solid* +300 mm, *dashed* −300 mm), *blue* for $z = \pm 900$ mm, (*solid* +900 mm, *dashed* −900 mm). The z position corresponds to the centre of the rotating coil probe. $z = 0$ corresponds to the longitudinal centre of the magnet

8.3 Summary

Rotating coils have been used for measuring the magnetic field of accelerator magnets. Then circular cylindrical multipoles have been used for describing the field. Based on these methods and the theory developed for rotating coils two new approaches were developed and demonstrated:

- the magnetic field of a dipole magnet was measured at different positions using a rotating coil. These measurements were then combined using a smooth weighting function, so that the field on the ellipse could be reconstructed and elliptical multipoles calculated. The circular multipoles were then calculated using a matrix whose coefficients were computed using analytic expressions. It was shown that the elliptical multipoles produce a better description of the field if more terms are used. For the S2LD the recalculated circular ones, however, were only representing the field well if not less than 7 or more than 29 multipole were used. Similar results were obtained for the other magnets, which show the advantage of the elliptical multipoles for describing fields within elliptical apertures.
- The sensitivity of a rotating coil to toroidal multipoles was calculated. The results show that the length of the coil probe has to be chosen carefully so that the artefacts created by errors in the geometric positioning of the coil probe is within acceptable limits.
- While circular cylindrical multipoles are appropriate to check and describe the field quality of a straight magnet, toroidal circular multipoles are required for describing the field quality of a curved particle beam in a dipole field. The inverse aspect ratio ε shows if the effects are significant and higher order perturbation terms have to be considered in field dynamics calculations.
- For machines with an inverse aspect ratio value ε and similar requirements to the accuracy of the field measurement as for SIS100 and SIS300 circular multipoles can still be used after these have been recalculated by the matrices (8.27) and (8.41). This also means that the perturbation terms of the toroidal cylindrical multipoles can be neglected. It shows why the classical approach of using cylindrical circular multipoles has been used successfully for beam dynamic calculations since decades.
- The same checks were also made for the NICA booster and collider machines. The evaluation showed that the size of the perturbation terms is nearly a factor of 4 larger than the ones for SIS100 and SIS300. Therefore it should be carefully evaluated if the perturbation terms have to be considered for beam dynamic calculations.

References

1. E. Fischer, P. Schnizer, A. Akishin, H. Khodzhibagiyan, A. Kovalenko, R. Kurnyshov, P. Shcherbakov, G. Sikler, W. Walter, Manufacturing of the first full size model of a SIS100 dipole magnet, in *WAMSDO Workshop*, (CERN, January 2009), pp. 147–156, number ISBN 978-92-9083-325-3

2. G. Sikler et al., *Full Size Model Manufacturing and Advanced Design Status of the SIS100 Main Magnets*, (WAMSDO at CERN, June 2008)
3. E. Fischer, H. Khodzhibagiyan, A. Kovalenko, P. Schnizer, Fast ramped superferric prototypes and conclusions for the final design of the SIS100 main magnets. IEEE Trans. Appl. Supercon. **19**(3), 1087–1091 (2009)
4. E. Fischer, P. Schnizer, R. Kurnyshov, B. Schnizer, P. Shcherbakov, Numerical analysis of the operation parameters of fast cycling superconducing magnets. IEEE Trans. Appl. Supercon. **19**(3), 1266–1267 (2009)
5. P. Schnizer, H.R. Kiesewetter, T. Mack, T. Knapp, F. Klos, M. Manderla, S. Rauch, M. Schöncker, R. Werkmann, Mole for measuring pulsed superconducting magnets. IEEE Trans. Appl. Supercon. **18**(2), 1648–1651 (2008)
6. E. Fischer, P. Schnizer, A. Mierau, P. Akishin, J. P. Meier, The SIS100 superconducting fast ramped dipole magnet, in *Proceedings of IPAC2014, Dresden, Germany* (2014)
7. E. Fischer, P. Schnizer, K. Sugita, J.P. Meier, A. Mierau, A. Bleile, P. Szwangruber, H. Müller, C. Roux, Fast ramped superconducting magnets for FAIR–production status and first test results. IEEE Trans. Appl. Supercon., **25**(3):1–5, (2015) Art.Nr: 4003805
8. P. Schnizer, B. Schnizer, P. Akishin, E. Fischer, Theory and application of plane elliptic multipoles for static magnetic fields. Nucl. Instrum. Methods Phys. Res., Sect. A. **607**(3), 505–516 (2009)
9. P. Schnizer, B. Schnizer, E. Fischer, Cylindrical circular and elliptical, toroidal circular and elliptical multipoles, fields, potentials and their measurement for accelerator magnet (Oct 2014), *arXiv preprint physics.acc-ph*
10. A.K. Jain, Harmonic coils, in *CAS Magnetic Measurement and Alignment*, ed. by S.Turner, (CERN, August 1998), pp. 175–217
11. P. Schnizer, *Measuring system qualification for LHC arc quadrupole magnets*. PhD thesis, TU Graz, (2002)
12. E. Fischer, P. Schnizer, A. Mierau, S. Wilfert, A. Bleile, P. Shcherbakov, C. Schroeder, Design and test status of the fast ramped superconducting SIS100 dipole magnet for FAIR. IEEE Trans. Appl. Supercon. **21**(3), 1844–1848 (2011)
13. G. Moritz, F. Klos, B. Langenbeck, Q. Youlun, K. Zweig, Measurements of the SIS magnets. IEEE Trans. Magn. **24**(2), 942–945 (1988)
14. C. Muehle, B. Langenbeck, A. Kalimov, F. Klos, G. Moritz, B. Schlitt, Magnets for the heavy-ion cancer therapy accelerator facility (HICAT) for the clinic in Heidelberg. IEEE Trans. Appl. Supercon. **14**(2), 461–464 (2004)
15. C. Priano, G. Bazzano, D. Bianculli, E. Bressi, I. De Cesaris, L. Vuffray, M. Pullia, M. Buzio, R. Chritin, D. Cornuet, J. Dutour, E. Froidefond, C. Sanelli, Magnetic modeling, measurements and sorting of the CNAO synchrotron dipoles and quadrupoles, in *Proceedings of IPAC'10, Kyoto, Japan*, (2010), pp. 280–282
16. J. Billan, L. Bottura, M. Buzio, G. D'Angelo, G. Deferne, O. Dunkel, P. Legrand, A. Rijl-lart, A. Siemko, P. Sievers, S. Schloss, L. Walckiers, Twin rotating coils for cold magnetic measurements of 15 m long LHC dipoles. IEEE T. Appl. Supercon. 1422–1446 (2000)
17. P. Schnizer, B. Schnizer, P. Akishin, E. Fischer, Plane elliptic or toroidal multipole expansions for static fields. Applications within the gap of straight and curved accelerator magnets. Int. J. Comput. Math. Electr. Eng. (COMPEL), **28**(4), (2009)
18. P. Schnizer, B. Schnizer, P. Akishin, E. Fischer, Toroidal circular and elliptic multipole expansions within the gap of curved accelerator magnets, in *14th International IGTE Symposium, Graz* (Technische Universität Graz, Austria, Institut für Grundlagen und Theorie der Elektrotechnik, 2010)
19. O. Brüning, P. Collier, P. Lebrun, S. Myers, R. Ostojic, J. Poole, P. Proudlock, *LHC Design Report* (CERN, Geneva, 2004)
20. E. Fischer, P. Schnizer, P. Akishin, R. Kurnyshov, A. Mierau, B. Schnizer, S.Y. Shim, P. Sherbakov, Superconducting SIS100 prototype magnets design, test results and final design issues. IEEE Trans. Appl. Supercon. **20**(3), 218–221 (2010)

21. FAIR–Facility for Antiprotons and Ion Research, Technical Design Report, Synchrotron SIS100 (Dec 2008)
22. H.G. Khodzhibagiyan, P.G. Akishin, A.V. Bychkov, A. Donyagin, A.R. Galimov, O.S. Kozlov, G.L. Kuznetsov, I.N. Meshkov, V.A. Mikhaylov, E.V. Muravieva, P.I. Nikitaev, A.V. Shabunov, A.V. Smirnov, A.Y. Starikov, G.V. Trubnikov, Status of the design and test of superconducting magnets for the NICA project, in *Proceedings of RUPAC2012, Saint-Petersburg, Russia*, (September 2012), pp. 149–151
23. H.G. Khodzhibagiyan, P.G. Akishin, A.V. Bychkov, A.D. Kovalenko, O.S. Kozlov, G.L. Kuznetsov, I.N. Meshkov, V.A. Mikhaylov, E.V. Muravieva, A.V. Shabunov, A.Y. Starikov, G.V. Trubnikov, Superconducting magnets for the NICA accelerator complex in Dubna. IEEE Trans. Appl. Supercon. **21**(3), 1795–1798 (June 2011)

Chapter 9
Error Propagation

In the preceding chapter two methods were presented which allow measuring elliptical cylindrical multipoles and toroidal circular multipoles based on rotating coil probes. Rotating coil probes are frequently used to measure the induced flux and then derive a set of coefficients of the cylindrical circular basis functions [1, 2]. The different sources of artefacts have been covered in literature (e.g. [3–6]).

Therefore the artefacts of the coil probe measurement itself are not reiterated but only the error propagation of the measured coefficients to the single set of cylindrical elliptical or toroidal circular harmonics is given here.

9.1 Error Propagation of Elliptic Multipoles Measurement

The following two artefacts sources are calculated:

1. The error propagation of the measurement error of an individual coil probe measurement $\Delta \mathbf{C}$ or
2. the artefacts created if one of the coil probes at the left or the right side are misplaced by a distance $\mathbf{d_z}$.

9.1.1 Description of Calculation Procedure

A method presented to combine three different coil probe measurements to a single set of cylindrical elliptic harmonics was given in Sect. 8.1.1 (see also Fig. 9.1). The treatment given in this section was found to be not straightforward to study the error propagation so a different approach is used here.

© Springer International Publishing AG 2017
P. Schnizer, *Advanced Multipoles for Accelerator Magnets*, Springer Tracts
in Modern Physics 277, DOI 10.1007/978-3-319-65666-3_9

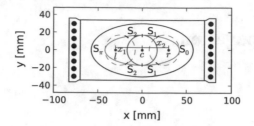

Fig. 9.1 Rotating coil probe positions for measuring elliptic multipole coefficients. The beam aperture is indicated by a *black ellipse*. The *green ellipse* indicates the ellipse used currently in reconstruction. The *blue circles* indicate the area covered by the coil probe. *l* denotes the centre of the coil probe for the measurement at the *"left"* side, *c* denotes the centre of the coil probe for the measurement at the centre and *r* denotes the centre of the coil probe for the measurement at the *right* side. S_0, S_1, S_2, S_π indicate the different intervals the ellipse is split up into for calculating the coefficients of the elliptic multipoles. x_1 gives the offset of the coil probe measurement from the centre to the *left* while x_2 gives the offset of the coil probe form the measurement to the *right*. The individual windings of the coil of the magnet are indicated by the *black filled circles* and its inner aperture by a line

1. As start a magnetic field is considered, which has been measured with a coil probe at three different positions. The coefficients of the elliptical cylindrical multipoles have then been derived using the procedure outlined in Sect. 8.1.
2. The accuracy of the coefficients obtained from the coil probe measurements has been assessed by methods as given in literature.
3. Now it is left over to study the error propagation. Here one considers that the elliptic multipoles are obtained by (4.29) from a measured field $\mathbf{B^m}(\psi)$. This measured field $\mathbf{B^m}(\psi)$ consists of

$$\mathbf{B^m}(\psi) = \mathbf{B}(\psi) + \Delta\mathbf{B}(\psi), \qquad (9.1)$$

with $\mathbf{B}(\psi)$ the real field and $\Delta\mathbf{B}(\psi)$ the field error, which is caused by measurement artefacts.

4. Thus the field represented by the artefacts of the measurements $\Delta\mathbf{B}(\psi)$ has to be calculated so that the error propagation can be derived. Here one can split it in three steps:

   (a) The contribution of each coefficient $\mathbf{C_n^{r,c,l}}$, which is obtained from a coil probe measurement, to the coefficient $\mathbf{E_m}$ is deduced.
   (b) Then the effect of a spurious harmonic of one of the measurements $\Delta\mathbf{C_n^{r,c,l}}$ on the coefficients of the elliptic multipoles $\Delta\mathbf{E_m}$ is calculated.
   (c) Thirdly the impact of a displaced rotating coil probe on the final measurement results is analysed using the "feed down effect" (see Sect. 4.1.2) for estimating the size of the individual spurious harmonics.

5. As elliptic multipoles are to be obtained using (4.29) the integral (4.29) is split up in the different intervals S_0, S_1, S_2, and S_π as given in Fig. 9.1.

6. The field in the different intervals is reconstructed by the measured coefficients of the circular harmonics obtained by the rotating coil probe.

Please note that for the interval S_0 or S_π only the coefficients are used, which are deduced form the coil probe measurement when it was placed at the right or left position. In the intervals S_1 and S_2 the field is reconstructed using the coefficients obtained from the right or left side and from the centre.

The calculation procedure is given in Sect. 9.1.2. The error propagation of the artefacts of the coil probe measurements $\Delta \mathbf{C}^{r,c,l}$ are given in Sect. 9.1.3. The effect of the displacements $\Delta \mathbf{z}^{r,c,l}$ is presented in Sect. 9.1.5.

9.1.2 Combining the Coefficients

The elliptic coefficients are obtained by a Fourier Transform (4.29) which is evaluated in Cartesian coordinates. Thus (4.29) is reformulated as

$$\mathbf{E_m} = \frac{1}{\pi} \int_{-\pi}^{\pi} \mathbf{B}(a \cos(\psi) + ib \sin(\psi)) \cos\left([m-1]\,\psi\right) d\psi. \qquad (9.2)$$

The field can be reconstructed from the coil probe measurement using

$$\mathbf{B}^{r,c,l}(\psi) = \sum_{n=1}^{N} \mathbf{C_n^{r,c,l}} \left(\frac{[a \cos \psi - x_0] + i\,[b \sin \psi]}{R_{\text{Ref}}} \right)^{n-1}. \qquad (9.3)$$

x_0 is then the offset of the coil probe from the centre of the ellipse. Thus it is set to

$$x_0 = x_m \qquad \qquad \text{for} \quad \mathbf{C_n^r} \qquad \qquad (9.4)$$
$$x_0 = 0 \qquad \qquad \text{for} \quad \mathbf{C_n^c} \qquad \qquad (9.5)$$
$$x_0 = -x_m \qquad \qquad \text{for} \quad \mathbf{C_n^l}. \qquad \qquad (9.6)$$

This reflects that the field reconstructed from the coil probe coefficients is not extrapolated too much from the area covered by the coil probe. The integral above could be split in 6 integrals for the different areas $S_0 \ldots S_\pi$ and solved by hand using binomial series expansion and the appropriate integrals of the sin cos products, as found in e.g. [7]. These expansions require to take several special cases into account.

The integral can be solved analytically up to any order; but on the other hand 20 elliptic coefficients are sufficient for the foreseen application. Thus the integrals were calculated symbolically for each n and m using Mathematica™. It was found that the integrals of Mathematica™ using the `Integrate` function gave singular terms, which were not straightforward to treat.

Using the reformulation of (4.29) the `FourierCoefficient` function can be used, which produced terms without singularities if it was evaluated for each individual term m and n.

To use this function not the interval of integration is split up but the interval in (9.2) is selected using `UnitBox` functions. These functions are defined by

$$U(x) = H(x^2 - 1/4) \tag{9.7}$$

with H, the Heaviside step function. Therefore $U(x)$

$$U(x) = \begin{cases} 1 \text{ for } -\frac{1}{2} \leq x \leq \frac{1}{2} \\ 0 \text{ for } x < -\frac{1}{2} \text{ or } x > \frac{1}{2} \end{cases}. \tag{9.8}$$

The different selection functions S are then defined by

$$S_0(\psi) = \frac{\psi}{2\psi_c}, \quad S_1(\psi) = -\frac{2\psi_c - 4([\psi \ \mathrm{sgn}(\psi)] \bmod \pi) + \pi}{2\pi - 4\psi_c}, \tag{9.9}$$

$$S_\pi(\psi) = \frac{(\psi \bmod (2\pi)) - \pi}{2\psi_c} \quad \text{and} \quad S_2(\psi) = \frac{-2\psi_c - 4([\psi \ \mathrm{sgn}(\psi)] \bmod \pi) + 3\pi}{2\pi - 4\psi_c}. \tag{9.10}$$

So S_0, S_1, S_2 and S_π selects the following part of the ellipse:

$$S_0: \quad -\psi_c \leq \psi \leq \psi_c \tag{9.11}$$

$$S_1: \quad -\frac{\pi}{2} \leq \psi \leq -\psi_c \quad \text{and} \quad \psi_c \leq \psi \leq \frac{\pi}{2} \tag{9.12}$$

$$S_2: \quad -\pi + \psi_c \leq \psi \leq -\frac{\pi}{2} \quad \text{and} \quad \frac{\pi}{2} \leq \psi \leq \pi - \psi_c \tag{9.13}$$

$$S_\pi: \quad -\pi \leq \psi \leq -\pi + \psi_c \quad \text{and} \quad \pi - \psi_c \leq \psi \leq \pi \tag{9.14}$$

The field **B** is integrated to obtain the coefficients E_m. This integral is split up in the different sections:

- the part S_0, where the field B^r is reconstructed using only the coil probe at the right side,
- the part S_π, where the field B^l is reconstructed using only the coil probe at the left side,
- and the parts S_1 and S_2, where the field B^c is constructed using the coefficients obtained by two measurements and a mixing polynomial (8.5). For the sections S_1 and S_2 the magnetic field is calculated using the coefficients obtained from the measurements at the centre and the coefficients deduced from the measurements at the left or the right side. This mixing polynomial is defined by

$$P(p) = 3p^2 - 2p^3, \tag{9.15}$$

with $0 \leq p \leq 1$.

Thus in the right apex of the magnet and therefore in the range $-\psi_c \leq \psi \leq \psi_c$ the field is deduced from the coefficients as obtained by the measurement at the right side:

$$\mathbf{B}^{\mathbf{p}}(\psi) = U(S_0(\psi))\mathbf{B}^{\mathbf{r}}(\psi). \tag{9.16}$$

Similarly in the left apex or in the interval $-\pi \leq \psi \leq -\psi_c$ and $\psi_c \leq \psi \leq \pi$ the field is given by

$$\mathbf{B}^{\mathbf{p}}(\psi) = U(S_\pi(\psi))\mathbf{B}^{\mathbf{l}}(\psi). \tag{9.17}$$

In the range of S_1 and S_2 the field is calculated using the coefficients obtained at the left or right side and the ones of the centre. S_1 and S_2 are discontinuous ranges. S_1 is adjacent to S_0 for the upper and lower part of the ellipse in the first and forth quadrant. S_2 is adjacent to S_π for the upper and lower part of the ellipse in the second and third quadrant.

In the range S_1 and thus of $-\pi/2 \leq \psi \leq -\psi_c$ and $\psi_c \leq \psi \leq \pi/2$ the field is reconstructed using the coefficients obtained by the measurement in the centre and at the right side. The field in this region is then given by

$$\mathbf{B}^{\mathbf{p}}(\psi) = \frac{1}{2} U(S_1(\psi)) \left[1 - P\left(S_1(\psi) + \frac{1}{2} \right) \right] \mathbf{B}^{\mathbf{r}}(\psi) +$$
$$\frac{1}{2} U(S_1(\psi)) \left[P\left(S_1(\psi) + \frac{1}{2} \right) \right] \mathbf{B}^{\mathbf{c}}(\psi). \tag{9.18}$$

In the same manner the field is reconstructed for S_2 for the interval $-\pi + \psi_c \leq \psi \leq -\pi/2$ and $\pi/2 \leq \psi \leq \pi - \psi_c$ by the coefficients obtained at the left and central position. Thus $\mathbf{B}^{\mathbf{p}}(\psi)$ is given by

$$\mathbf{B}^{\mathbf{p}}(\psi) = \frac{1}{2} U(S_2(\psi)) \left[P\left(S_2(\psi) + \frac{1}{2} \right) \right] \mathbf{B}^{\mathbf{c}}(\psi) +$$
$$\frac{1}{2} U(S_2(\psi)) \left[1 - P\left(S_2(\psi) + \frac{1}{2} \right) \right] \mathbf{B}^{\mathbf{l}}(\psi). \tag{9.19}$$

Combining the expressions (9.16)–(9.19)

$$\mathbf{B}^{\mathbf{p}}(\psi) = \left\{ U(S_0) + \frac{1}{2} U(S_1) \left[1 - P\left(S_1 + \frac{1}{2} \right) \right] \right\} \mathbf{B}^{\mathbf{r}}(\psi)$$
$$+ \frac{1}{2} \left\{ U(S_1) \left[P\left(S_1 + \frac{1}{2} \right) \right] + U(S_2) \left[P\left(S_2 + \frac{1}{2} \right) \right] \right\} \mathbf{B}^{\mathbf{c}}(\psi)$$
$$+ \left\{ U(S_\pi) + \frac{1}{2} U(S_2) \left[1 - P\left(S_2 + \frac{1}{2} \right) \right] \right\} \mathbf{B}^{\mathbf{l}}(\psi), \tag{9.20}$$

Table 9.1 Dimensions of the geometry of the measurement of the CSLD. a, b...half axes of the ellipse, R_c...largest radius of the coil probe, R_{Ref}...reference radius

Parameter	Value	Unit
a	45	mm
b	17	mm
R_c	17	mm
R_{Ref}	40	mm
$\left\| \Delta \mathbf{C^{r,c,l}} / \mathbf{C^l} \right\|$	$< 10^{-5}$ @ R_c	
$\left\| \Delta \mathbf{z} \right\|$	3	mm

is obtained. S_0, S_1, S_2 and S_π are functions of ψ. Using (9.3) and (9.20) the integral (9.2) can be expressed by

$$\mathbf{E}_m = \frac{1}{\pi} \int_{-\pi}^{\pi} \mathbf{B^P}(\psi) \, \cos\left([m-1]\,\psi\right) \, d\psi. \qquad (9.21)$$

To study the error propagation of the measured coil probe the terms of the obtained integral (9.20) has to be related to the different measured coefficients $\mathbf{C_n^{r,c,l}}$. The integrand $\mathbf{B^P}$ is a sum and so the integral can be split in its different parts. As the integral can be subdivided, the coefficient $\mathbf{E_m}$ can be seen as a sum of

$$\mathbf{E_m} = \mathbf{E_m^r} + \mathbf{E_m^c} + \mathbf{E_m^l}, \qquad (9.22)$$

with $\mathbf{E_m^{r,c,l}}$ with the elliptic harmonic as obtained from the right, central, or left interval. Then the terms of the integrals can be sorted into different matrices $M^{r,c,l}$ which are defined by

$$\mathbf{E_m^{r,c,l}} = M_{m,n}^{r,c,l} \, \mathbf{C_n^{r,c,l}}. \qquad (9.23)$$

These matrices $M_{m,n}^{r,c,l}$ can now be used to study the error propagation of the errors of the coefficients $\Delta \mathbf{C_n^{r,c,l}}$.

9.1.3   Error Propagation of the Measured Coefficients

The terms of $M_{m,n}^{r,c,l}$ contain only terms of polynomials in a, b, R_c or functions of ψ_c. The expressions of $M_{m,n}^{r,c,l}$ are lengthy and not repeated here. Instead the value of coefficients are given for the measurement geometry (see Table 9.1). ψ_c is the value of the angular variable, where the right circle intersects the ellipse, thus $\psi_c \approx 64.93$ degree. The values are then

$$M_{m,n}^r \approx \frac{1}{10\,000} \begin{pmatrix} 4304 & 2977 & 1764 & 3721 & 1335 & 2897 & 3141 \\ 3092 & 3104 & 2300 & 2846 & 2347 & 2488 & 2863 \\ 662 & 2934 & 3225 & 1554 & 3771 & 2236 & 2406 \\ -805 & 1765 & 3265 & 1804 & 3401 & 3146 & 2591 \\ -565 & 102 & 1982 & 3137 & 1986 & 4080 & 3726 \\ 259 & -761 & 268 & 3262 & 1848 & 3502 & 4821 \\ 430 & -394 & -634 & 1361 & 2923 & 2217 & 4498 \\ -14 & 298 & -412 & -734 & 2779 & 2083 & 2963 \\ -284 & 365 & 177 & -1076 & 590 & 2709 & 2008 \\ -91 & -60 & 324 & -22 & -1369 & 1956 & 2397 \end{pmatrix} \qquad (9.24)$$

for the contribution of right coil probe and

$$M_{m,n}^c \approx \frac{1}{10\,000} \begin{pmatrix} 1393 & 0 & -1463 & 0 & -76 & 0 & 878 \\ 0 & 129 & 0 & -306 & 0 & 70 & 0 \\ -1323 & 0 & 1418 & 0 & -48 & 0 & -757 \\ 0 & -357 & 0 & 852 & 0 & -240 & 0 \\ 1131 & 0 & -1288 & 0 & 376 & 0 & 417 \\ 0 & 502 & 0 & -1221 & 0 & 476 & 0 \\ -860 & 0 & 1092 & 0 & -787 & 0 & 75 \\ 0 & -540 & 0 & 1355 & 0 & -771 & 0 \\ 568 & 0 & -858 & 0 & 1136 & 0 & -619 \\ 0 & 479 & 0 & -1259 & 0 & 1066 & 0 \end{pmatrix} \qquad (9.25)$$

for the central coil probe. The contribution of the left coil probe is given by

$$M_{m,n}^l = (-1)^{m+n} M_{m,n}^r. \qquad (9.26)$$

These matrices already show:

1. that all coefficients are real. Thus the measured coefficients of the normal harmonics contribute only to the coefficients of the normal elliptic harmonics and the measured coefficients of the skew harmonics only contribute to the coefficients of the elliptic skew harmonics,
2. that the coefficients for the higher harmonics are mainly obtained from the measured higher order harmonics,
3. that only even harmonics are measured by the central coil probe,
4. that all harmonics contribute to the main harmonic. At first sight it may seem odd that the higher order harmonics contribute so much to the term E_1. But the higher order harmonics are by a factor of $1/1000\ldots1/10000$ smaller than the main harmonic. So their contribution is only small for the fields of accelerator magnets.
5. As the values of the elements of the first column are large, the measured coefficient $C_1^{r,c,l}$ contributes considerably to all E_m. Therefore the procedure to correct artefacts of the dipole field measurement has to be executed with care.

The equations above show that a linear transformation of the coefficients obtained by the coil probe measurements to the coefficients of the elliptic harmonics can be made. Inserting (9.23) in (9.22) gives

$$\mathbf{E_m} = M^r_{m,n} \, \mathbf{C^r_n} + M^c_{m,n} \, \mathbf{C^c_n} + (-1)^{m+n} \, M^r_{m,n} \, \mathbf{C^l_n} \, . \tag{9.27}$$

Using the orthogonality of $\cos(m\psi)$ inserting (9.1) in (9.2) yields

$$\mathbf{E^m_m} = \mathbf{E_m} + \Delta \mathbf{E_m}, \tag{9.28}$$

with $\mathbf{E^m_m}$ the coefficients of the elliptic multipoles deduced by measurement, $\mathbf{E_m}$ the coefficients of the real field and $\Delta \mathbf{E_m}$ the coefficients created by the artefacts. An estimate of the error propagation of the $\Delta \mathbf{C^{r,c,l}_n}$ to the $\Delta \mathbf{E_m}$ is derived; in particular the absolute value $|\Delta \mathbf{E_m}|$ is estimated, which is given by

$$|\Delta \mathbf{E_m}| \geq \left| \mathbf{E^m_m} - \mathbf{E_m} \right| . \tag{9.29}$$

The matrices $M^{l,c,r}_{m,n}$ reflect the geometry and thus are error free. Then the relative errors of the elliptic coefficients can be estimated by

$$|\Delta \mathbf{E_m}| = \sqrt{\sum_{n=1}^{N} \left[\left(|M^r_{m,n}| \, |\Delta \mathbf{C^l_n}| \right)^2 + \left(|M^c_{m,n}| \, |\Delta \mathbf{C^c_n}| \right)^2 + \left(|M^r_{m,n}| \, |\Delta \mathbf{C^r_n}| \right)^2 \right]},$$
$$m = 1 \ldots M, \quad (9.30)$$

assuming that the $\Delta \mathbf{C^{r,c,l}_n}$ are uncorrelated. While the above equation is generally valid, for the measurement of the CSLD dipole $n = 1 \ldots N$, $N = 10$, and $M = 10$ was used. Even if the same coil probe equipment is used for all three measurements, the measurement artefacts are not systematic as typically spurious harmonics are similar in amplitude but not necessarily of the same phase. The term $(-1)^{m+n}$ has been removed as one can not assume that the measurement error of the left and the right coil probe cancel. Using the above equations $\Delta \mathbf{E_m}$ can be estimated if the $\Delta \mathbf{C^{r,c,l}_n}$ are known.

Equation (9.30) is lengthy and thus studied for three different cases, to facilitate its understanding. If one assumes that the measurement accuracy of all coefficients sets, obtained by the coil probe, are of similar accuracy, (9.30) can be rewritten to

$$|\Delta \mathbf{e^{\varsigma}_m}| = \sqrt{\sum_{n=1}^{N} \left(|M^r_{m,n}|^2 + |M^c_{m,n}|^2 + |M^r_{m,n}|^2 \right) |\Delta \mathbf{c^{\varsigma}_n}|} , \tag{9.31}$$

with $\Delta \mathbf{e^{\varsigma}_m} = \Delta \mathbf{E_m}/\mathbf{E_1}$ and $\Delta \mathbf{c^{\varsigma}_n} = \Delta \mathbf{C_n}/\mathbf{C_1}$. If one considers that the relative accuracy of all higher order harmonics is the same then the equation above can be further simplified to

$$|\Delta \mathbf{e_m^{\varsigma}}| = \sqrt{\sum_{n=1}^{N} \left(|M_{m,n}^r|^2 + |M_{m,n}^c|^2 + |M_{m,n}^r|^2 \right)} \, |\Delta \mathbf{c}^{\varsigma}| \, . \qquad (9.32)$$

For the following estimates $\mathbf{E_m}$ is assumed to be 1 T. Setting all $|\Delta \mathbf{c}^{\varsigma}| = 10^{-4} \, T$ at the coil radius R_c (9.32) yields for $m = 1 \dots 10$

$$|\Delta \mathbf{e_m^{\varsigma}}| \approx \left(1.57 \;\; 1.3 \;\; 1.55 \;\; 1.54 \;\; 1.83 \;\; 1.8 \;\; 1.94 \;\; 2.12 \;\; 2.24 \;\; 1.97 \right) \qquad (9.33)$$

and setting $|\Delta \mathbf{c}^{\varsigma}| = 10^{-5} \, T$ at R_c gives for $m = 1 \dots 10$

$$|\Delta \mathbf{e_m^{\varsigma}}| \approx \left(0.16 \;\; 0.13 \;\; 0.16 \;\; 0.15 \;\; 0.18 \;\; 0.18 \;\; 0.19 \;\; 0.21 \;\; 0.22 \;\; 0.2 \right) \, . \qquad (9.34)$$

For rotating coil probes with compensation arrays the coefficients of the higher order harmonics are measured much more precisely than the coefficient of the main harmonic. For a typical coil probe, which has a compensation array designed for measuring the field of a dipole, the $|\Delta \mathbf{c_n^{\varsigma}}|$ are typically

$$|\Delta \mathbf{c_1^{\varsigma}}| = 5 \cdot 10^{-4} \, T \;\; \text{at} \;\; R_c \quad \text{and} \quad |\Delta \mathbf{c_{n \neq 1}^{\varsigma}}| = 10^{-5} \, T \;\; \text{at} \;\; R_c \, , \qquad (9.35)$$

with R_c the largest radius of the coil probe. Inserting it into (9.31) yields for $m = 1 \dots 10$

$$|\Delta \mathbf{e_m^{\varsigma}}| \approx \left(7.85 \;\; 0.13 \;\; 0.16 \;\; 0.15 \;\; 0.18 \;\; 0.18 \;\; 0.19 \;\; 0.21 \;\; 0.22 \;\; 0.2 \right) \, . \qquad (9.36)$$

This shows that $\mathbf{C_1}$ is best obtained directly from the measured coefficient $\mathbf{C_1^c}$, while the errors of all $\mathbf{E_m}$ for $m \geq 2$ are of similar magnitude.

Equation (9.31) has more than one minimum, thus it was evaluated for different values of ψ_c. It was found that for the sets of $|\Delta \mathbf{c_n^{\varsigma}}|$ as given above, the error propagation is at a practical minimum value if ψ_c is chosen such, that it matches the point were the ellipse and the right circle intersect.

9.1.4   Error Propagation to Coefficients of the Circular Multipoles

The calculated elliptic multipoles are recalculated to circular multipoles using (4.46). Therefore

$$\mathbf{F_s} = T_{sm} \, \mathbf{E_m}. \qquad (9.37)$$

The letter $\mathbf{F}$ was chosen for the coefficients of *common* circular multipoles so that these are not confused with the multipole coefficients obtained by the measurement. Given that (9.22)

$$\mathbf{E_m} = \mathbf{E_m^r} + \mathbf{E_m^c} + \mathbf{E_m^l}$$

one can also write

$$\mathbf{F_s^{r,c,l}} = T_{sm}\,\mathbf{E_m^{r,c,l}}. \tag{9.38}$$

Please note that the $\mathbf{F^{r,c,l}}$ are coefficients for the central coordinate system $x = 0$, $y = 0$. Thus the multipole coefficient $\mathbf{F_s^{r,c,l}}$ of the cylindrical circular multipoles obtained from $\mathbf{E_m^{r,c,l}}$ are given by

$$\mathbf{F_s^{r,c,l}} = T_{sm}\,M_{mn}^{r,c,l}\,\mathbf{C_n^{r,c,l}} \tag{9.39}$$

and thus

$$\mathbf{F} = \mathbf{F^r} + \mathbf{F^c} + \mathbf{F^l}. \tag{9.40}$$

The $\mathbf{C_m^{r,c,l}}$ at R_c are mapped to $\mathbf{F_s^{r,c,l}}$ at $R_{\mathrm{Ref}} = 40$ mm by

$$T_{sm}M_{mn}^r \approx \begin{pmatrix} 0.4304 & 0.2977 & 0.1764 & 0.3721 & 0.1335 & 0.2897 & 0.3141 \\ 0.3092 & 0.3104 & 0.23 & 0.2846 & 0.2347 & 0.2488 & 0.2863 \\ 0.0662 & 0.2934 & 0.3225 & 0.1554 & 0.3771 & 0.2236 & 0.2406 \\ -0.0805 & 0.1765 & 0.3265 & 0.1804 & 0.3401 & 0.3146 & 0.2591 \\ -0.0565 & 0.0102 & 0.1982 & 0.3137 & 0.1986 & 0.408 & 0.3726 \\ 0.0259 & -0.0761 & 0.0268 & 0.3262 & 0.1848 & 0.3502 & 0.4821 \\ 0.043 & -0.0394 & -0.0634 & 0.1361 & 0.2923 & 0.2217 & 0.4498 \end{pmatrix} \tag{9.41}$$

for the coefficients as obtained from the right measurement and

$$T_{sm}M_{mn}^c \approx \begin{pmatrix} 0.1564 & 0. & -0.1713 & 0. & 0.0224 & 0. & 0.0755 \\ 0. & 0.2097 & 0. & -0.5156 & 0. & 0.2351 & 0. \\ -0.9526 & 0. & 1.212 & 0. & -0.8899 & 0. & 0.1764 \\ 0. & -0.8996 & 0. & 2.2665 & 0. & -1.3589 & 0. \\ 1.8444 & 0. & -2.7341 & 0. & 3.4793 & 0. & -2.0031 \\ 0. & 1.554 & 0. & -3.9915 & 0. & 2.8377 & 0. \\ -2.0429 & 0. & 3.4418 & 0. & -5.7475 & 0. & 4.1479 \end{pmatrix} \tag{9.42}$$

for the coefficients from the central measurement. The first row of these matrices (9.41) and (9.42) shows that adjusting the measured dipole coefficient, so that the field is continuous in the overlap area, is a critical step for achieving the required measurement accuracy, as otherwise all harmonics were mainly influenced by the dipole. The coefficients of the columns at the right of these matrices, and in particular of $T_{sm}M_{mn}^r$ show that the value of the higher order harmonics of C_m will be strongly dependent on measurement artefacts as the values of the coefficients to the right get larger and larger, in particular due to the products (4.44) and (4.45) in (4.46). Therefore coefficients whose amplitude are close to the limit of the coil probe measurement system must be discarded as these will significantly deteriorate the obtained coefficients.

The error propagation can be calculated as above by

$$|\Delta \mathbf{F}_s^{r,c,l}| = \sqrt{\sum_{n=1}^{N} \left(|T_{sm} M_{mn}^{r,c,l}| \, |\Delta \mathbf{C}_n^{r,c,l}| \right)^2}.$$ (9.43)

All matrix elements are similar in magnitude, and thus an approximation is not straight forward. Setting $\mathbf{C}_1^c$ to 1 T and assuming a measurement accuracy of $\Delta \mathbf{C}_m^{r,c,l}$

$$|\Delta \mathbf{c}_1^s| = 5 \cdot 10^{-4} \, T \text{ at } R_c \quad \text{and} \quad |\Delta \mathbf{c}_{n \neq 1}^s| = 10^{-5} \, T \text{ at } R_c.$$ (9.44)

one obtains for $s = 1 \dots 10$

$$|\Delta \mathbf{F}_s^r| \approx \frac{1}{10\,000} \left(3.6 \; 0.2 \; 0.5 \; 0.8 \; 1.5 \; 1.9 \; 2.5 \; 2.3 \; 2.4 \; 1. \; 1. \right)$$ (9.45)

and

$$|\Delta \mathbf{F}_s^c| \approx \frac{1}{10\,000} \left(1.3 \; 0. \; 0.2 \; 0.3 \; 0.6 \; 0.6 \; 1. \; 0.5 \; 0.8 \; 0.1 \; 0.2 \right)$$ (9.46)

for the error propagation for the right side and the centre side. The error propagation of the artefacts of the measurement at the left side gives the same contribution as the error propagation of the artefacts at the right side due to (9.26). Hence the total error is then given by

$$|\Delta \mathbf{F}_s| = \sqrt{|\Delta \mathbf{F}_s^r|^2 + |\Delta \mathbf{F}_s^c|^2 + |\Delta \mathbf{F}_s^r|^2} =$$ (9.47)

$$= \frac{1}{10\,000} \left(5.3 \; 0.2 \; 0.8 \; 1.2 \; 2.2 \; 2.8 \; 3.6 \; 3.3 \; 3.5 \; 1.5 \; 1.5 \right).$$ (9.48)

One can see that the error of the coefficient of the dipole term is of similar accuracy as if it is deduced from the measurement at the centre. The error of all higher order multipoles is considerable larger than $\Delta \mathbf{C}_n^{r,c,l}$. But these are the errors at the radius of the coil probe R_c, which is 17 mm, while the $\Delta \mathbf{F}$ are for R_{Ref}, which is 40 mm.

As comparison one simply scales the measurement error of the measurement at the centre $\Delta \mathbf{C}_n^c = \Delta \mathbf{C}_n^s$ from the coil radius R_c to the reference radius R_{Ref} using

$$\frac{|\Delta \mathbf{C}_n^c|}{|\mathbf{C}_1^c|} \left(\frac{R_{\text{Ref}}}{R_c} \right)^n = \frac{|\Delta \mathbf{C}_n^c|}{|\mathbf{C}_1^c|} \left(\frac{40}{17} \right)^n$$ (9.49)

$$\approx \frac{1}{10\,000} \left(5 \; 0.2 \; 0.6 \; 1.3 \; 3.1 \; 7.2 \; 17. \; 39.9 \; 93.9 \; 221.1 \right).$$

These values roughly show which field artefact each measured coefficient generates at the reference radius, if only the central coil probe were used. One can see, that the artefact created by the lower order terms are similar to the accuracy of the central coil probe, while for the higher order terms a significant accuracy gain is reached if the combined coefficients are used.

The $\left|\Delta e_m^c\right|$ are roughly a factor 5–10 smaller than $|\Delta F_s|$. Therefore a field inter-
polation based on elliptic multipoles will give a considerably better approximation
of the real field than an interpolation based on the circular multipoles.

9.1.5 Influence of Coil Probe Displacement

The error propagation of spurious coefficients $\Delta C_n^{r,c,l}$ has been discussed above.
The effect of a misplaced coil probe can thus be derived by deducing which spurious
coefficients $\Delta \mathbf{C}_n^{r,c,l}$ are generated by a misplaced coil probe. The coefficients obtained
from a rotating coil probe measurement are valid in the coordinate system of rotation;
the artefacts of a coil probe, which is misplaced from its ideal measurement location
by $\mathbf{dz}$, has to be treated by translating the coordinate system and recalculating the
coefficients by (4.9). This equation can be expressed in matrix notation by (8.34)

$$\mathbf{C}'_n = \underbrace{\left[\mathcal{L}_{kn} \binom{k-1}{n-1} \left(\frac{\mathbf{d}'_z}{R_{\mathrm{Ref}}} \right)^{k-n} \right]}_{\mathcal{L}_{kn}^{dr}} \mathbf{C}'_k. \tag{9.50}$$

If the displacement $\Delta \mathbf{z}$ can not be measured by some other means and thus its effect
can not be corrected, it creates spurious harmonics of the strength

$$|\Delta \mathbf{C}_n| = \left| \mathcal{L}_{kn}^{dz} - I_{kn} \right| |\mathbf{C}'_k|, \tag{9.51}$$

with I the identity matrix. Their effect on the measured elliptic harmonics can be
derived from (9.30).

9.2 Toroidal Multipole Measurement

It was shown in Sects. 8.2.3 and 8.2.4 that only the term $\mathcal{L}_L$ gives significant contributions for SIS100 or NICA. $\mathcal{L}^L$ is given by (8.37)

$$\mathcal{L}^L = \frac{L^2}{3 R_{\mathrm{Ref}}^2} \mathcal{L}^{ddr}.$$

The term $\mathcal{L}^{ddr}$ depends on a coil probe displacement $\mathbf{dz}$. Its effect has been discussed
in Sect. 8.2.3. So only the error propagation of ΔL of $\mathcal{L}^L$ is to be considered and
thus G (8.41) can be approximated as a complex variable as the not analytic terms
in (8.41) can be neglected. Then $|\Delta \mathbf{G}|$ is given by

$$|\Delta \mathbf{G}| = \varepsilon \frac{2L}{3R_{\text{Ref}}^2} \mathcal{L}^{ddr} \, \Delta L \, |\mathbf{C_n}|. \tag{9.52}$$

The coils of the coil probes are precisely machined to better than $0.5\,$mm. Thus $\Delta L/L$ is typically smaller than 0.1% (see Table 5.1 for the used coil probe parameters). If one evaluates (9.52) for $\mathbf{C_n}$ of 5 units, and $\mathbf{d_z}$ of 3 mm, one obtains for $|\Delta \mathbf{G}|$

$$|\Delta \mathbf{G}| \approx \frac{1}{10\,000} \left(0. \ 0.39 \ 0.79 \ 1.24 \ 1.76 \ 2.41 \ 3.19 \ 4.17 \ 5.38 \ 6.86 \right) . \tag{9.53}$$

This means that the dipole field component is not affected by this error. All higher order harmonics are distorted by less than 0.1%. Therefore the measurement accuracy due to the coil probe length uncertainty will not significantly attribute to the total measurement error (9.55) and can be neglected. The extra feed down effect due to the curvature of the magnet will increase the given feed down effect. The error propagation of $\mathcal{L}^L$ due to an uncertainty of $\mathbf{d_z}$ was treated in Sect. 8.2.3.

9.3 Measurement Accuracy Estimate for the CSLD

The descriptions derived above gave an estimate for the measurement error $\Delta \mathbf{E}$ if spurious coefficients $\Delta \mathbf{C^{r,c,l}}$ were present in the measurement. In this section now it is evaluated which spurious harmonics are created by a displaced measurement coil probe of some unknown $\left| \Delta z^{l,r} \right|$ (see also Table 9.1).

The seven meters long movable anti-cryostat (see Sect. 6.2) hardly attained that accuracy, but rather an accuracy of $\left| \Delta z^{l,r} \right| \approx 5\,$mm. Assuming a measured field with coefficients

$$|\mathbf{C_n}| = \frac{1}{10\,000} \left(10\,000 \ \tfrac{1}{10} \ 2 \ \tfrac{1}{10} \ 1 \ \tfrac{1}{10} \ 1 \ \tfrac{1}{10} \ 1 \ \tfrac{1}{10} \right) \left(\frac{17}{40} \right)^{n-1} \tag{9.54}$$

(9.51) yields for $\left| \Delta z^{l,r} \right| = 5\,$mm

$$|\Delta \mathbf{C_n}| = \frac{1}{10\,000} \left(0.09 \ 0.22 \ 0.04 \ 0.03 \ 0.02 \ 0.01 \ 0 \ 0 \ 0 \ 0 \right) . \tag{9.55}$$

Thus the quadrupole is affected considerably but not the higher order multipoles.
So one can conclude that

- To improve the measurement accuracy the uncertainty in the position of the coil probes shall be controlled to be better than $\approx 2\,$mm.

- The measurement accuracy for coefficients obtained with the coil probe shall be increased to 0.05 units (or 50 ppm).
- These improvements will allow measuring the elliptic multipoles with sufficient accuracy. As the coefficients of the circular multipoles are obtained from the coefficients of elliptical ones by a transform matrix (4.46), the uncertainties of the coefficients of the circular harmonics are larger than the ones of the elliptical cylindrical harmonics (see Sect. 9.1.4).

At the currently achieved measurement accuracy for elliptical multipoles $\mathbf{E_m}$ the magnetic field is approximated with an error of ≈ 0.5 units within the elliptic boundary.

Typically the relative accuracy of the coefficient $|\Delta \mathbf{C_n}/\mathbf{C_1}|$, which has been obtained by the measurement, is 0.1 units. These errors add up and thus especially the accuracy of the lower order coefficients will be diminished. It also shows that it is better to neglect small coefficients which are close to the measurement accuracy.

9.4 Summary

The error propagation analysis revealed that the coefficients of the elliptic multipoles can be deduced with an accuracy of ≈ 0.2 units for $m \geq 2$. Furthermore the interpolated field homogeneity will be predicted with an accuracy of ≈ 0.5 units.

As the coefficients of the circular multipoles are derived from the elliptical ones by matrix multiplication, and thus several elliptical coefficients contribute to one circular one, their accuracy will be considerably less than for the elliptic multipoles. This shows once again that the elliptic multipoles are more adapted to this elliptical problem.

For applications with higher accuracy demands the following steps allow improving the measurement results:

- an improved accuracy of the coefficients obtained by the coil probes
- a more precise positioning of the coil probe or
- using coil probes with a larger measurement radius.

References

1. W. Weber, Über die Anwendung der magnetischen Induktion auf Messung der Inclination mit dem Magnetometer. Ann. der Physik **166**, 209–247 (1853)
2. J.K. Cobb, R.S. Cole, Spectroscopy of quadrupole magnet, in *Proceedings/International Symposium on Magnet Technology (MT-1)*, (Stanford, USA, 1965), pp. 431–446
3. W.G. Davies, The theory of the measurement of magnetic multipole fields with rotating coil magnetometers. Nucl. Instrum. Methods. Phys. Res. A: Accel. Spectrom. Detect. Assoc. Equip. **311**(3), 399–436 (1992)

4. A.K. Jain, Harmonic coils, in *CAS Magnetic Measurement and Alignment* ed. by S. Turner (CERN, Aug 1998), pp. 175–217
5. A. Devred, M. Traveria, Magnetic field and flux of particle accelerator magnets in complex formalism, *not published*, 1994
6. P. Schnizer, *Measuring system qualification for LHC arc quadrupole magnets*. PhD thesis, TU Graz, 2002
7. S.I. Gradshteyn, I.M. Ryzhik, *Table of Integrals* (Academic Press, Series and Products, 1965)

Chapter 10
Conclusions

This treatise covered the following topics:

- different geometries required to model the magnetic field of accelerator magnets in advanced multipoles: cylindrical elliptic multipoles, toroidal circular multipoles and toroidal elliptic multipoles,
- their relation to cylindrical circular multipoles,
- the used test station and measurement setup,
- calculation of these multipoles based on numerical data obtained from FEM calculations and the quality of these descriptions and
- the measurement of these advanced multipoles on built and measured prototype magnets for the SIS100 of the FAIR project.

The original motivation for these developments was to produce a sound basis for describing and measuring the magnetic field of the SIS100 magnets. But the descriptions and deductions are generally applicable to machines of similar geometry and characteristic parameters.

The magnetic field in vacuum and air is described by linear material approximations. Therefore the Laplace equation describes its properties (see Sect. 2.1.2 page 13). The following geometries (see Chap. 3 page 21) were presented (see Chap. 4 page 35) along with their multipole descriptions:

1. cylindrical circular multipoles
2. cylindrical elliptic multipoles
3. local toroidal circular multipoles and
4. local toroidal elliptic multipoles.

The first two are exact solutions of the potential equation while the last two are approximations with solutions linear in ε (which is the ratio of the reference radius to the major radius of the torus $\varepsilon = R_{Ref}/R_C$). Cylindrical circular multipoles have been the work horse for describing the magnetic field properties in beam dynamics. The

© Springer International Publishing AG 2017
P. Schnizer, *Advanced Multipoles for Accelerator Magnets*, Springer Tracts in Modern Physics 277, DOI 10.1007/978-3-319-65666-3_10

magnetic field B_x, B_y represented with cylindrical elliptic coordinates and multipoles can be recalculated to circular cylindrical multipoles (see Sect. 4.2.2 page 47). Both these descriptions are analytic and can be calculated using complex functions.

The potential equation is solved in the toroidal geometries by transforming the differential equation to the corresponding cylindrical one with an additional term depending on ε^2 or $\bar{\varepsilon}^2$, which is neglected for the applications given here. The solutions for the cylindrical differential equations are known and are found in textbooks. The potential of the toroidal circular multipoles, however, is not a complex analytic function as it contains at least one term which is not analytic. Yet it was shown that its potential and field can be described using complex functions. Due to the non analytic term different field expansions have to be used for the normal and skew multipole coefficients.

The applications (Chap. 7) showed that cylindrical elliptical multipoles can be used to describe the magnetic field within an elliptical domain and that these can be recalculated to cylindrical ones (see Sect. 7.2 page xxx). The results obtained on measurements, however, showed that this translation can have some shortcoming: a set of coefficients, so that the series can not be cut at some arbitrary point (see Sect. 8.1 page xxx and in particular Fig. 8.7 page xxx). For the cylindrical elliptical multipoles the first coefficients are the largest ones while for the transformed cylindrical circular coefficients "a band" of alternating coefficients of large size is found. If the series is cut within this band, the reconstructed field inhomogeneity will be much larger than the originally measured field inhomogeneity. Thus the elliptical multipole coefficients provide a better field representation than the transformed circular ones. Beam dynamic calculations can profit from the improved description using the conversion of elliptical to cylindrical multipoles. But the use of the circular multipoles can have the aforementioned short comings which the direct use of the elliptical multipoles has not.

The sensitivity of a rotating coil probe in a curved magnet was studied based on toroidal circular multipoles (see Sect. 8.2 page 116). A criterion was given to estimate the maximum length of the rotating coil probe based on the reference and curvature radius of the machine. Furthermore it was deduced that for SIS100 it is sufficient just to recalculate cylindrical circular harmonics by the given matrices. Other machines with a larger ε will have to evaluate if the terms dependent on ε are still small enough so that these can be neglected.

Based on the measurement accuracy of the rotating coil probe system the error propagation of the measured coefficients to the final combined coefficients was derived and the current accuracy limit of these combined coefficients was calculated (Chap. 9).

The measurement of these advanced multipoles will advance beam dynamic studies if also these advanced field descriptions are implemented within the codes. Concise and simple formulations were given. The elliptical ones use functions, whose evaluation is computationally more expensive; but this is easily mitigated by the increased computer power. The advanced multipoles also show why the standard approach of circular multipoles works so well even for non cylindrical apertures as

- the cylindrical circular multipoles are a simple polynomial term,
- elliptical multipoles can be translated to circular ones and
- the terms dependent in ε need only to be taken into account for machines with a larger ε and still more precise field description requirements as SIS100.

The developments summarised here all supported the design and development of the SIS100 prototype and first of series magnets. The coefficients obtained using the elliptical multipoles were used to study the beam dynamic properties of SIS100. The SIS100 prototype magnets were measured on the test facility at GSI. Based on the procedures described within this treatise, coefficients valid within the whole elliptical aperture, were obtained and used for evaluating the transverse beam dynamics. Currently all series SIS100 magnets are being procured, built, measured and sorted to the locations in the machine. The design and measurement analysis is based on the results given in this treatise and thus the accomplishments are already applied during the realisation phase of SIS100.

10.1   Outlook

This document covers different set of multipoles, their application and measurement. The results summarised above show already next steps to pursue.

- The basis functions of the cylindrical elliptical multipoles were derived from their mapping to the cylindrical circular multipoles and further extended to elliptical multipoles on elliptical coordinates. This allows showing that only the ones given are required for describing a magnetic field of Cartesian components.

For toroidal elliptical coordinates the approximation of the differential equation was presented. A full treatment will be an interesting enhancement.

- It was shown that rotating coils can be used for measuring toroidal circular multipoles. During this treatment the artefacts, which a horizontal or vertical offset creates, were given as well. These achievements have to be extended to all the measurement artefacts, which have been considered for rotating coils when used to measure cylindrical circular multipoles: e.g. rotation of the coil probes axis around the vertical axis (y-axis).

All these results are based on definite integrals evaluated with Mathematica™ for the first 20 terms. Solving the indefinite integral will allow proving the deduced equation by differentiation.

- Toroidal elliptical multipoles are not required for describing the field of SIS100 magnets with the demanded resolution; these will have to be used for machines with a bigger $\bar{\varepsilon}$ and similar or more challenging requirements to the description of the field. Then one has to check if the methods described for measuring the cylindrical elliptical multipoles can be adjusted to obtain toroidal elliptical multipoles.

- The impact of mechanical artefacts on the measurement quality of search coils and flux meters can be derived using toroidal circular multipoles.

Appendix A
Changes to Previous Publications

This chapter is ment for readers who followed previous publications or even based work of their own on these texts.

The following changes were made to the results given in previous publications in comparison to the presentation in this document:

European Convention   The papers describing the theoretical development used the US convention as it is more close to standard Fourier Series. European convention (see (2.42) p. 26) is used throughout this text (original papers [1–7]).

Toroidal multipoles   The complex representation of the basis function and the calculation of the real terms was changed. Here concise descriptions of the basis functions are given. The representation chosen now allows describing the basis functions of the potential and the field as complex functions. However these are not analytic and different functions have to be used for the "normal" and "skew" coefficients (original papers [7]).

Measuring toroidal multipoles   The theory for measuring toroidal circular multipoles with a rotating coil probe was outlined in [7]; there the area normal of the coil probe was not taken properly into account. This is corrected in this treatise.

© Springer International Publishing AG 2017

P. Schnizer, *Advanced Multipoles for Accelerator Magnets*, Springer Tracts in Modern Physics 277, DOI 10.1007/978-3-319-65666-3

Appendix B
Mathematica Scripts

B.1 Coil Probe Within a Torus

The scripts, listed below, were used to calculate the field which a rotating coil probe measures inside a torus. The field is described with local toroidal cylindrical multi-poles as given in (4.122), p. 63.

The calculations of the integrals of the measurement coil probe and the matrices to interpret these measurements were presented in Sect. 8.2, p. 116. The following script was used to evaluate the integral and the next one to check that the matrix prediction was correct. The scripts were split as the integral evaluation required approximately 3 hours on contemporary hardware with up to 16 processes evaluating the integrals in parallel.

evaluate_integrals_parallel.m

```
(* ::Package:: *)


(* ================================================= *)
(* The local toroidal basis functions *)
TmComplexCommonTerm[zs_, eps_, n_] :=
    (zs)^(n − 1) * (1 − 1 / 2 * eps * Re[zs]);


TmComplexTerm[zs_, eps_, n_] :=
Module[{common, notAnalytic, result},
    common = TmComplexCommonTerm[zs, eps, n];
    notAnalytic = − eps  / (2 * n) *
        {Im[zs^(n − 1) * zs] *I, Re[zs^(n − 1) * zs]};
    result  = common + notAnalytic;
    result
];


(* ================================================= *)
(* helper functions *)
variabelsAndCoil[expr_] :=
    Block[{subs, result},
        subs = (r/R * Exp[I * theta] + dx/R + dy/R * I +
```

© Springer International Publishing AG 2017
P. Schnizer, *Advanced Multipoles for Accelerator Magnets*, Springer Tracts
in Modern Physics 277, DOI 10.1007/978-3-319-65666-3

```
                eps * z^2 / R^2);
    result = ReplaceAll[expr,   zs ->  subs]* Exp[I * theta];
    result
]

integrateOverCoil[expr_] :=
    Block[{realVars},
    realsVars={R, r, r1, r2, theta, z, L, eps, dy, dx};
    Integrate[expr, {z, 0, L},{r, r1, r2},
        Assumptions->Element[realVars, Reals] &&
        R > 0 && eps >0 && r2 > r1]
]

buildTerms[expr_, start_Integer, end_Integer, normalOrSkew_] :=
 Block[{termNormal, fluxNormal, flx},
    termNormal = expr;
    fluxNormal = variabelsAndCoil[termNormal];
    flx = Table[fluxNormal, {n, start, end}];
    flx = ComplexExpand[Re[flx * normalOrSkew]];
    flx = ReplaceAll[flx, onlyLinearInEps];
    flx
    ]

(* The integrals *)
(* ══════════════════════════════════════════════════ *)
minterm = 1
maxterm = 21
onlyLinearInEps = Table[eps^n -> 0,
    {n, 2, 2 * Max[minterm, maxterm]}];
savename = ("pierres_integral_"
    <> ToString[minterm] <> "_" <> ToString[maxterm])
Print[savename];

basisterms = TmComplexTerm[zs, eps, n]

(* The normal components *)
flxn = buildTerms[basisterms[[1]], minterm, maxterm, 1];

(* The skew components *)
flxs = buildTerms[basisterms[[2]], minterm, maxterm, I];

(* Revert the list ... faster parallel execution .... *)
flxn = flxn[[-1;;1;;-1]];
flxs = flxs[[-1;;1;;-1]];
(* Prepare to evaluate all at once *)
allflx = Apply[Join, Transpose[{flxn, flxs}]];
dims = Dimensions[allflx];
If[Length[dims] == 1, Null,
    Print["List of integrands not one dimensional!", dims];
    Exit[]];
ninteg = Length[allflx];
Print["Starting integral evaluation ", ninteg, " integrands"];
{usedTime, integralall} =  AbsoluteTiming[
```

```
      ParallelMap[integrateOverCoil, allflx]
];
Print["Finished: required ", N[usedTime,1], "seconds"];

(* separate the result again ... *)
(*    normal *)
integraln = integralall[[1;; ;; 2]]
(*    skew *)
integrals = integralall[[2;; ;; 2]]
integraln = integraln[[-1;;1;;-1]]];
integrals = integrals[[-1;;1;;-1]]];

(* and store the result for later analysis *)
Save[savename, {integraln, integrals}];
```

check_integrals_all1.m

```
(* ::Package:: *)

(*
 * Check if the guessed representation of the integrals matches the
 * integrals achieved with mathematica
 *)
(* ============================================= *)
On[Assert];
(* ============================================= *)
(*
 * The matrix U has to be evaluated to more one colum than any
 * other matrix here. This function adds a zero colume to a given
 * matrix
 *)
addColumnToMat[mat_] :=
Block[{dims, nrows, tcol, result},
    dims = Dimensions[mat];
    Assert[Length[dims] == 2];
    nrows = dims[[1]];
    tcol = Array[Function[{trow}, 0], nrows];
    result = Append[Transpose[mat], tcol];
    result = Transpose[result];
    result
]
(* for convienience ... *)
identityMatrix[nterms_Integer] :=
    addColumnToMat[IdentityMatrix[nterms]]

checkZeroInt[val_] := (MatchQ[val, _Integer] && val == 0)

checkZeroVector[mat_] :=
Module[
    {result, nrows, trow, dims,tmp },
    dims = Dimensions[mat];
    nrows = dims[[1]];
    (* check every value if it is an integer *)
    tmp= Table[checkZeroInt[Part[mat, trow]],{trow, nrows}];
```

```mathematica
    (* now lets see if all were True! *)
    result = Apply[And,Map[Apply[And, #]&, tmp]];
    result
]                      .


createFeedDownMatrix[dzs_, n_Integer, default_:0] := Module[
    {feeddowntable, result},
    feeddowntable = createLowerMat[
        Function[{trow, tcol},
            Binomial[trow−1,tcol−1] *(dzs)^(trow − tcol)],
        n, n, Null, default
    ];
    result = feeddowntable;
    result
]


createLowerMat[f_, nrows_Integer, ncols_Integer, args_, default_:0] :=
Module[
    {trow, tcol, result},
    result = Array[
        Function[{trow, tcol},
            If[trow >= tcol, f[trow, tcol, args] , default]
            ],
        {nrows, ncols}
    ];
    result
]


(*========================================================*)
Get["pierres_integral_1_21"];
integralnt = integraln;
integralst = integrals;
Dimensions[integralnt]
(* replacement lists to get from coil sensitity to R2 and R1 etc *)
invcoilSensitivities=Table[
    sk[n]−>(r2^n− r1^n)* L/R^(n−1)/n,{n, 1, Length[integralnt]*2}];


createFeedDownMatrixPaper[dzs_, n_Integer, default_:0] :=
    createFeedDownMatrix[dzs, n, default] − IdentityMatrix[n];


(* The different sub matrices *)
(* The most difficult term  *)
ConstructR2MatCheck[nterms_] :=
Module[
{trow, tcol, nrows, ncols, lowermat, feeddownmat, feeddownscale, dims,
    upperdiagonal, corrections, result, coeffsreal, coeffsupper,
    coeffsimag, coeffsscale},
nrows = nterms;
ncols = nterms;
dims = {nrows, ncols};
(* The part similar scaling with dz / R *)
feeddownmat =  createFeedDownMatrixPaper[dx/R + I * dy/R, nterms] +
 IdentityMatrix[nterms];
```

```
feeddownscale = Array[Function[{trow, tcol},  1/trow], {nrows, ncols}];
result = feeddownscale * feeddownmat / 4 ;

(* now comes this peculiar part in dx and dy .... *)
(* the coefficients *)
coeffsreal =  createLowerMat[
        Function[{trow, tcol},  trow * tcol / (trow − tcol + 1)],
    nrows, ncols, Null];

coeffsreal = (coeffsreal +
    Array[Function[{trow, tcol}, KroneckerDelta[tcol, 1 ]], dims]);

coeffsscale = createLowerMat[
        Function[{trow, tcol},(2 − tcol + 2 * trow) / (tcol) ],
    nrows, ncols, Null];
coeffsscale += Array[Function[{trow, tcol},
    − trow * KroneckerDelta[tcol, 1]],dims];

corrections =  coeffsreal  ( dy / R −coeffsscale* dx /R * I) ;
result = result * corrections ;
result = Expand[result];
addColumnToMat[result]
]
```

```
Ldr [nterms_Integer] := Expand[
    addColumnToMat[
    createFeedDownMatrixPaper[(dx/R + I * dy/R), nterms, 0]
    ]
]

U[nterms_Integer] :=
 Array[Function[{trow, tcol},
    KroneckerDelta[trow + 1, tcol] * (trow +1) / ((trow) * 4 )],
    {nterms, nterms+1}
]

createFeedDownDerivMatrix[dzs_, n_Integer, default_:0] :=
    Block[
    {fdDzs, dfdDzs, tdzs},
        fdDzs = createFeedDownMatrixPaper[tdzs, n, default];
        dfdDzs = D[fdDzs, tdzs];
        dfdDzs = ReplaceAll[dfdDzs, tdzs −> dzs];
        addColumnToMat[dfdDzs]
]

LL[nterms_Integer] :=
    Block[
    {result, dfdDzs},
    dfdDzs = createFeedDownDerivMatrix[dx/R + dy/R*I, nterms, 0];
    result = (L^2 / (R^2  * 3)) (dfdDzs);
    result = Expand[result];
    result
]
```

```
Lsk[nterms_Integer] :=
Block[{result, scale, dfdDzs},
    scale = createLowerMat[
        Function[{trow, tcol, unused},
                    1/4 * 1/(tcol+1) *  sk[tcol+2]/sk[tcol]],
        nterms, nterms+1, Null];
    dfdDzs = createFeedDownDerivMatrix[dx/R + dy/R*I, nterms, 0];
    result = scale *  dfdDzs;
    result = Expand[result];
    result
]

LR2[nterms_Integer] := ConstructR2MatCheck[nterms]

LR20[nterms_Integer] := addColumnToMat[
  Array[
    Function[{trow, tcol},
        KroneckerDelta[tcol, 1] * 1 /( 2 * trow )* ((dx + I * dy)/R)^(trow)],
    {nterms, nterms}]
]

toroidalNormalToCoilNormal[ndim_] :=
(identityMatrix[ndim]+Re[Ldr[ndim]]
  +eps(−U[ndim]+Re[LL[ndim]]−Re[Lsk[ndim]]+Im[LR2[ndim]]+Re[LR20[ndim]]))

toroidalNormalToCoilSkew[ndim_] :=
(                     +Im[Ldr[ndim]]
    +eps(          +Im[LL[ndim]]−Im[Lsk[ndim]]−Re[LR2[ndim]]))

toroidalSkewToCoilNormal[ndim_] :=
(                     −Im[Ldr[ndim]]
    +eps(          −Im[LL[ndim]]+Im[Lsk[ndim]]+Re[LR2[ndim]]−Im[LR20[ndim]]))

toroidalSkewToCoilSkew[ndim_] :=
(identityMatrix[ndim]+Re[Ldr[ndim]]
    + eps(−U[ndim]+Re[LL[ndim]]−Re[Lsk[ndim]]+Im[LR2[ndim]]));

constructIntegrals[mat_, term_] :=
Block[{tmp, nterms, dims, termtable, result},
    dims = Dimensions[mat];
    Assert[Length[dims] == 2];
    nterms = dims[[2]];
    termtable = Table[term * sk[n], {n, 1, nterms}];
    tmp = mat. termtable;
    tmp = ReplaceAll[tmp, invcoilSensitivities];
    result = Expand[tmp];
    result
    ]

calculateNDim[mat_]:= Max[Dimensions[mat]]
selectOnlyFirstRows[mat_, dims_] :=
Block[{result, tdims, nrows, ncols},
```

```
    Assert[Length[dims] == 2];
    tdims = Dimensions[mat];
    Assert[Length[tdims] == 2];
    nrows = dims[[1]];
    ncols = dims[[2]];
    Assert[nrows <= tdims[[1]]];
    Assert[ncols <= tdims[[2]]];
    result = Part[mat, ;; nrows, ;; ncols];
    result;
]

ndim = calculateNDim[integralnt];
Print["Building normal component terms up to term ", ndim];
refdim = Dimensions[integralnt];
convnn = constructIntegrals[
    ComplexExpand[toroidalNormalToCoilNormal[ndim]], Cos[n * theta]];
(*
*   the terms in Sin have to be taken times -1 as
* Phi = ComplexExpand[Re[(Bn + I * An) * Exp[I * n * theta]]]
* Phi = Bn Cos[n theta]-An Sin[n theta]
*)
convns = constructIntegrals[
    ComplexExpand[toroidalNormalToCoilSkew[ndim]],      Sin[n * theta]];
convns = convns * -1;
Print[Dimensions[convnn]];
Print["Comparing normal component terms to integrals"];
result = Expand[TrigReduce[integralnt]] - convnn - convns;
Print["   Simplifying Diff"];
result = Simplify[result];
checkn = checkZeroVector[result];
If[checkn == True,
    Print["Integral for normal components fully represented (expected!)"],
    Print["Representing the integral for the normal components left over:",
    result, checkn]
];
Print["Building skew component terms"];
ndim = calculateNDim[integralnt];
convsn = constructIntegrals[
    ComplexExpand[toroidalSkewToCoilNormal[ndim]], Cos[n * theta]];
convss = constructIntegrals[
    ComplexExpand[toroidalSkewToCoilSkew[ndim]], Sin[n * theta]];
(*
* again ... the coefficients have to be multiplied -1 as the normal
* coil probesees the field - Sin[n * theta]
*)
convss = convss * -1;
Print[Dimensions[convsn]];
Print["Comparing skew component terms to integrals"];
result = Expand[TrigReduce[integralst]] - convsn - convss;
Print["   Simplifying Diff"];
result = Simplify[result];
checks = checkZeroVector[result];
If[checks == True,
```

```
Ldr [nterms_Integer] := Expand[
    addColumnToMat[
    createFeedDownMatrixPaper[(dx/R + I * dy/R), nterms, 0]
    ]
]

U[nterms_Integer] :=
 Array[Function[{trow, tcol},
    KroneckerDelta[trow + 1, tcol] * (trow +1) / ((trow) * 4 )],
    {nterms, nterms+1}
]

createFeedDownDerivMatrix[dzs_, n_Integer, default_:0] :=
    Block[
    {fdDzs, dfdDzs, tdzs},
        fdDzs = createFeedDownMatrixPaper[tdzs, n, default];
        dfdDzs = D[fdDzs, tdzs];
        dfdDzs = ReplaceAll[dfdDzs, tdzs -> dzs];
        addColumnToMat[dfdDzs]
]

LL[nterms_Integer] :=
    Block[
    {result, dfdDzs},
    dfdDzs = createFeedDownDerivMatrix[dx/R + dy/R*I, nterms, 0];
    result = (L^2 / (R^2 * 3)) (dfdDzs);
    result = Expand[result];
    result
]

Lsk[nterms_Integer] :=
Block[{result, scale, dfdDzs},
    scale = createLowerMat[
        Function[{trow, tcol, unused},
                        1/4 * 1/(tcol+1) * sk[tcol+2]/sk[tcol]],
        nterms, nterms+1, Null];
    dfdDzs = createFeedDownDerivMatrix[dx/R + dy/R*I, nterms, 0];
    result = scale * dfdDzs;
    result = Expand[result];
    result
]

LR2[nterms_Integer] := ConstructR2MatCheck[nterms]

LR20[nterms_Integer] := addColumnToMat[
  Array[
    Function[{trow, tcol},
      KroneckerDelta[tcol, 1] * 1 /( 2 * trow )* ((dx + I * dy)/R)^(trow)],
    {nterms, nterms}]
]

toroidalNormalToCoilNormal[ndim_] :=
(identityMatrix[ndim]+Re[Ldr[ndim]]
```

```
   +eps(−U[ndim]+Re[LL[ndim]]−Re[Lsk[ndim]]+Im[LR2[ndim]]+Re[LR20[ndim]]))

toroidalNormalToCoilSkew[ndim_]  :=
(                    +Im[Ldr[ndim]]
   +eps(            +Im[LL[ndim]]−Im[Lsk[ndim]]−Re[LR2[ndim]]))

toroidalSkewToCoilNormal[ndim_]  :=
(                    −Im[Ldr[ndim]]
   +eps(          −Im[LL[ndim]]+Im[Lsk[ndim]]+Re[LR2[ndim]]−Im[LR20[ndim]]))

toroidalSkewToCoilSkew[ndim_]  :=
(identityMatrix[ndim]+Re[Ldr[ndim]]
   + eps(−U[ndim]+Re[LL[ndim]]−Re[Lsk[ndim]]+Im[LR2[ndim]]));

constructIntegrals[mat_, term_] :=
Block[{tmp, nterms, dims, termtable, result},
    dims = Dimensions[mat];
    Assert[Length[dims] == 2];
    nterms = dims[[2]];
    termtable = Table[term * sk[n], {n, 1, nterms}];
    tmp = mat. termtable;
    tmp = ReplaceAll[tmp, invcoilSensitivities];
    result = Expand[tmp];
    result
    ]

calculateNDim[mat_]:= Max[Dimensions[mat]]
selectOnlyFirstRows[mat_, dims_]  :=
Block[{result, tdims, nrows, ncols},
    Assert[Length[dims] == 2];
    tdims = Dimensions[mat];
    Assert[Length[tdims] == 2];
    nrows = dims[[1]];
    ncols = dims[[2]];
    Assert[nrows <= tdims[[1]]];
    Assert[ncols <= tdims[[2]]];
    result = Part[mat, ;; nrows, ;; ncols];
    result;
]

ndim = calculateNDim[integralnt];
Print["Building normal component terms up to term ", ndim];
refdim = Dimensions[integralnt];
convnn = constructIntegrals[
    ComplexExpand[toroidalNormalToCoilNormal[ndim]], Cos[n * theta]];
(*
*    the terms in Sin have to be taken times −1 as
* Phi = ComplexExpand[Re[(Bn + I * An) * Exp[I * n * theta]]]
* Phi = Bn Cos[n theta]−An Sin[n theta]
*)
convns = constructIntegrals[
    ComplexExpand[toroidalNormalToCoilSkew[ndim]],        Sin[n * theta]];
convns = convns * −1;
```

```
Print[Dimensions[convnn]];
Print["Comparing normal component terms to integrals"];
result = Expand[TrigReduce[integralnt]] - convnn - convns;
Print["   Simplifying Diff"];
result = Simplify[result];
checkn = checkZeroVector[result];
If[checkn == True,
    Print["Integral for normal components fully represented (expected!)"],
    Print["Representing the integral for the normal components left over:",
        result, checkn]
];
Print["Building skew component terms"];
ndim = calculateNDim[integralnt];
convsn = constructIntegrals[
    ComplexExpand[toroidalSkewToCoilNormal[ndim]], Cos[n * theta]];
convss = constructIntegrals[
    ComplexExpand[toroidalSkewToCoilSkew[ndim]], Sin[n * theta]];
(*
 * again ... the coefficients have to be multiplied -1 as the normal
 * coil probesees the field - Sin[n * theta]
 *)
convss = convss * -1;
Print[Dimensions[convsn]];
Print["Comparing skew component terms to integrals"];
result = Expand[TrigReduce[integralst]] - convsn - convss;
Print["   Simplifying Diff"];
result = Simplify[result];
checks = checkZeroVector[result];
If[checks == True,
```

```
    Print["Integral for skew components fully represented (expected!)"],
    Print["Representing the integral for the normal components left over:",
        result, checks]
];

Print["Reached end of Programm"];
```

Appendix C
Approximate Inversion of a Perturbed Matrix

A matrix C is considered, which can be split into two matrices A and B by

$$C = A + eB. \tag{C.1}$$

It is assumed that the inverse of the matrix A is known and that

$$|eB| \ll |A| \tag{C.2}$$

so that eB can be seen as perturbation of the matrix A. In this case the inverse C^{-1} is given by

$$C^{-1} = (A + eB)^{-1} = A^{-1} - eA^{-1}BA^{-1} + \mathcal{O}\left(e^2\right). \tag{C.3}$$

This can be proven by

$$(A + eB)(A + eB)^{-1} = (A + eB)\left(A^{-1} - eA^{-1}BA^{-1} + \mathcal{O}\left(e^2\right)\right) \tag{C.4}$$

$$= AA^{-1} + e(BA^{-1} - AA^{-1}BA^{-1}) + \mathcal{O}\left(e^2\right) \tag{C.5}$$

$$= E + e(BA^{-1} - EBA^{-1}) + \mathcal{O}\left(e^2\right) \tag{C.6}$$

$$= E + \mathcal{O}\left(e^2\right). \tag{C.7}$$

Similarly

$$(A + eB)^{-1}(A + eB) = \left(A^{-1} - eA^{-1}BA^{-1} + \mathcal{O}\left(e^2\right)\right)(A + eB) \tag{C.8}$$

$$= A^{-1}A + e\left(-A^{-1}BA^{-1}A + A^{-1}B\right) + \mathcal{O}\left(e^2\right) \tag{C.9}$$

$$= E + e\left(-A^{-1}BE + A^{-1}B\right) + \mathcal{O}\left(e^2\right) \tag{C.10}$$

$$= E + \mathcal{O}\left(e^2\right). \tag{C.11}$$

© Springer International Publishing AG 2017
P. Schnizer, *Advanced Multipoles for Accelerator Magnets*, Springer Tracts in Modern Physics 277, DOI 10.1007/978-3-319-65666-3

References

1. P. Schnizer, B. Schnizer, P. Akishin, E. Fischer, Field representation for elliptic apertures. Technical report, Gesellschaft für Schwerionenforschung mbH, Planckstraße 1, D-64291 Darmstadt, February 2007
2. P. Schnizer, B. Schnizer, P. Akishin, E. Fischer, Magnetic field analysis for superferric accelerator magnets using elliptic multipoles and its advantages. IEEE T. Appl. Supercon. **18**(2), 1605–1608 (2008)
3. P. Schnizer, B. Schnizer, P. Akishin, E. Fischer, Theoretical field analysis for superferric accelerator magnets using plane elliptic or toroidal multipoles and its advantages, in *The 11th European Particle Accelerator Conference*, June 2008, pp. 1773–1775
4. P. Schnizer, B. Schnizer, P. Akishin, E. Fischer, Field representation for elliptic apertures. Technical report, Gesellschaft für Schwerionenforschung mbH, Planckstraße 1, D-64291 Darmstadt, January 2008
5. P. Schnizer, B. Schnizer, P. Akishin, E. Fischer, Theory and application of plane elliptic multipoles for static magnetic fields. Nucl. Instrum. Methods Phys. Res. Sect. A: Accel. Spectrometers, Detectors Assoc. Equip. **607**(3), 505–516 (2009)
6. P. Schnizer, B. Schnizer, P. Akishin, E. Fischer, Toroidal circular and elliptic multipole expansions within the gap of curved accelerator magnets, in *14th International IGTE Symposium, Graz, Institut für Grundlagen und Theorie der Elektrotechnik* (Technische Universität Graz, Austria, September 2010)
7. P. Schnizer, B. Schnizer, P. Akishin, E. Fischer, Plane elliptic or toroidal multipole expansions for static fields. Applications within the gap of straight and curved accelerator magnets. Int. J. Comput. Math. Electr. Eng. (COMPEL) **28**(4) (2009)

Printed in the United States
By Bookmasters